Cambridge astrophysics series

T0207054

X-ray detectors in astronomy

In this series

X-RAY DETECTORS
IN ASTRONOMY

G. W. FRASER

*Lecturer, X-ray Astronomy Group, Department of Physics,
University of Leicester*

CAMBRIDGE UNIVERSITY PRESS

Cambridge

New York   New Rochelle

Melbourne   Sydney

CAMBRIDGE UNIVERSITY PRESS
Cambridge, New York, Melbourne, Madrid, Cape Town, Singapore, São Paulo, Delhi

Cambridge University Press
The Edinburgh Building, Cambridge CB2 8RU, UK

Published in the United States of America by Cambridge University Press, New York

www.cambridge.org
Information on this title: www.cambridge.org/9780521106030

First published 1989
This digitally printed version 2009

A catalogue record for this publication is available from the British Library

Library of Congress Cataloguing in Publication data
Fraser, G. W.
X-ray detectors in astronomy / G. W. Fraser.
p.   cm. – (Cambridge astrophysics series)
Bibliography: p.
Includes index.
ISBN 0-521-32663-X
1. X-ray astronomy – Measurement.   2. X-ray astronomy – Instruments.
3. Photon detectors.   I. Title.   II. Series.
QB472.F73   1989
522′.686–dc19   88-21417   CIP

ISBN 978-0-521-32663-6 hardback
ISBN 978-0-521-10603-0 paperback

Contents

Preface

The first cosmic X-ray source was discovered in June, 1962, during the flight of an Aerobee sounding rocket from the White Sands missile range in New Mexico (Giacconi *et al.*, 1962). As the rocket spun about its axis, three small gas-filled detectors scanned across a powerful source of low-energy X-rays in the constellation of Scorpius, in the southern sky. Even though the position of the source (later designated Sco X-1) could only be determined to within an area of some hundred square degrees, cosmic X-ray astronomy had begun.

As usually recounted, the story of Sco X-1 and the birth of X-ray astronomy bears a not inconsiderable resemblance to the story of X-rays themselves. The element of serendipity seems all-important in both discoveries. Wilhelm Roentgen, in 1896, had been intent on measuring the aether waves emitted by a low-pressure gas discharge tube when, by chance, he discovered his new and penetrating radiation. In 1962, the expressed aim of the American Science and Engineering (AS&E)–MIT research group led by Riccardo Giacconi was to detect the X-ray emission, not of distant stars, but from the moon.

Detailed consideration undermines this neat parallel. There is in fact a clear evolutionary line linking X-ray astronomy and the pioneering solar studies carried out in the USA by the Naval Research Laboratory (NRL) group under Herbert Friedman. Friedman's solar X-ray observations had begun with the flight of a captured German V2 rocket in September, 1949. As early as 1956, the NRL group had detected a hint of the cosmic X-ray background radiation whose formal 'discovery' accompanied that of Sco X-1, and by 1957 were flying detectors specifically to search for cosmic X-ray sources (Friedman, 1972). Nor was an awareness of the X-ray sky confined to the USA. Boyd (1979) has recalled a 1959 minute of the British National Committee on Space Research which read: 'Current theories suggest that there may be objects in the sky with strong X-ray emission, although inconspicuous visually. A search for these objects is a matter of great interest and importance.'

Recent first-hand testimony (Tucker and Giacconi, 1985) makes it plain that cosmic X-ray astronomy was born, not really by accident, but out of the determination of the AS&E researchers to beat down the sensitivities of the then available X-ray detectors until the first faint stellar signals, extrapolated from the known X-ray luminosity of the sun, emerged from the instrumental noise. The search for lunar X-ray fluorescence was a sideshow, a means of securing funds for astrophysical X-ray detector development in moon-obsessed, post-Sputnik America. Fortune favoured the AS&E group only in making the intrinsic X-ray luminosity of Sco X-1 so enormously, and unexpectedly, greater than that of the sun.

X-ray astronomy has taken many giant strides since 1962. X-ray sources up to seven orders of magnitude fainter than Sco X-1 have now been located with positional uncertainties of only a few arcseconds. In adolescence and maturity, as at birth, progress in the subject has been intimately linked to advances in photon-counting electronic X-ray detectors. It seems timely to describe the first 25 years of astronomical X-ray instrumentation and to summarise areas of current detector research, giving particular emphasis to imaging devices and to non-dispersive devices of high spectral resolution. Perhaps because it is a true space astronomy (getting above the earth's absorbing atmosphere is not just highly desirable but absolutely essential), X-ray astronomy illustrates supremely well the maxim: 'Astronomy advances only as fast as its instrumentation allows'.

In 25 years X-ray astronomy has become part of the mainstream of astrophysics. Modern astronomers bring data from different regions of the electromagnetic spectrum to bear on the study of particular objects, so that the interpretation of X-ray data is no longer confined to specialists with an assured first-hand knowledge of the photon-collecting hardware. An up-to-date, comprehensive account of detection techniques would therefore appear to be of interest, not only to the X-ray hardware specialist, in the context of a review, but to the wider astronomical community.

X-ray astronomy is, moreover, linked by its instrumentation to fields as diverse as particle physics, medicine, X-ray diffraction studies, X-ray microscopy and fusion plasma diagnostics. A number of recent conferences, recognising these links, have tried to bring together X-ray astronomers and workers in these other fields. A very important aim of this monograph is to stress the broad applicability of new detector technologies developed for astronomical X-ray research.

In providing an oblique history of the subject through the development of its hardware, this monograph may lastly prove to be of interest to the general, astronomically minded reader with a background in physics. Appendix A lists a number of reviews which would allow the reader to trace the same history through the astrophysics of the X-ray sources.

The timescales of satellite projects in X-ray astronomy are now very long. Ten years passed between approval of the EXOSAT Observatory by the European Space Agency (ESA) and launch of the spacecraft in May, 1983. The gestation periods of future missions, such as the US Advanced X-ray Astrophysics Facility (AXAF) and the European X-ray spectroscopy 'cornerstone' mission (XMM) promise to be even longer. Where does the 'frontier' lie in such a subject? Of course it must be with the data analysts, poring over the latest images and spectra from the satellites now in orbit. On the other hand, these people are using instruments whose designs were frozen long ago, instruments which may be a decade out of date. The frontier of X-ray astronomy also lies in those labs around the world where, amid cries of frustration and blown preamplifiers, the next generation of X-ray detector is being developed. It is to sufferers at this frontier that this monograph is dedicated.

Leicester, January, 1988

Acknowledgements

The author wishes to thank the following colleagues for their comments on the manuscript: H. D. Thomas, M. R. Sims, J. F. Pearson, D. H. Lumb and M. J. L. Turner. Thanks are also due to the members of the XMM Instrument Working Group (Chairman Tony Peacock, ESTEC), during whose debates many points of detector design became clearer to the author, and to those many colleagues worldwide who gave their consent for the reproduction of figure material. Finally, sincere thanks to Professor Ken Pounds (Director, University of Leicester X-ray Astronomy Group) for initiating and encouraging this project.

Units and constants

Most of this monograph is concerned with X-ray energies in the range $0.1 < E < 50\,\text{keV}$, with regular excursions outside the range as the detector technology demands. One kiloelectronvolt (1 keV) equals 1.6×10^{-16} joules. We shall, however, use X-ray energy E and X-ray wavelength λ interchangeably. If λ is expressed in ångströms $(1\,\text{Å} = 10^{-10}\,\text{m})$ and E is in keV then:

$$E = 12.4/\lambda$$

The following fundamental physical constants appear in the text:

permittivity of free space $\varepsilon = 8.85 \times 10^{-12}$ farads/metre (F/m)
Boltzmann's constant $k = 1.38 \times 10^{-23}$ joules/kelvin (J/K)
electron charge $e = 1.60 \times 10^{-19}$ coulombs (C)
electron charge-to-mass ratio $e/m = 1.76 \times 10^{11}$ coulombs/kilogram (C/kg).

I

Observational techniques in X-ray astronomy

1.1 Instrumental sensitivity

From the viewpoint of its instrumentation, X-ray astronomy possesses a certain moral simplicity. It is a perpetual battle of good versus evil; that is, of signal versus noise.

The observation of a weak point source of X-ray flux F (photons/cm^2 (detector area) s keV (detector bandwidth)) must always be made in the presence of an unwanted background B (counts/cm^2 s keV). This may be regarded as the sum of the intrinsic detector background B_i (arising, usually, from the very complex interaction of the near-earth radiation environment with the detection medium) and the diffuse X-ray sky background B_d, discovered on the same sounding rocket flight as Sco X-1 (Giacconi et al., 1962). If the quantum detection efficiency of the detector is Q counts/photon and its aperture is Ω steradians (sr), then

$$B_d = Q\Omega j_d$$

where j_d is the diffuse background flux in photons/cm^2 s keV sr.

If the observed quantities are all constant during the measurement time t, statistical fluctuations in the background B determine the sensitivity of the detector. The minimum detectable flux, F_{min}, for a given signal-to-noise ratio S is that flux which produces a count S standard deviations of B above its mean. Assuming that the bandwidth of the instrument is δE, the geometric area for the collection of source photons (and diffuse background) is A_s and the geometric area for detector background is A_b, one may show that:

$$F_{min} = (S/QA_s)\{B_i A_b + Q\Omega j_d A_s)/t\delta E\}^{\frac{1}{2}} \tag{1.1}$$

Choosing a value of S is equivalent to selecting a confidence level for the source detection. For example, $S=3$ corresponds to 99.8% confidence.

Equation (1.1) is a fundamental relationship, which remains valid down to a sensitivity limit determined by source confusion, when it becomes statistically likely that two or more sources of equal brightness are simultaneously present in each instrumental resolution element. It may be elaborated to include the real energy-dependence of terms such as F, j_d and Q. Limiting forms of the sensitivity equation can be derived to cover special observing modes, such as diffuse background dominance, measurement from a rotating or precessing platform, and so on (Peterson, 1975). Similar equations can be constructed for the detection of X-ray lines once a broad-band detection has been made. Implicit in the derivation of all such equations is the assumption that source detection is made on a photon-by-photon basis. Principally because of the weakness of the X-ray sources, integrating detectors such as X-ray film have never had a place in cosmic (as opposed to solar) X-ray astronomy. A flux of one photon per square centimetre per second (1–10 keV) observed at the earth constitutes a rather bright cosmic X-ray source. The 'standard candle' of X-ray astronomy – the Crab Nebula – contributes ~ 3 photons/cm^{2}s in this energy band.

In this chapter, we shall consider the evolution of the instrumentation used in X-ray astronomy, using eq. (1.1) as our guide.

1.2   Early days

The discovery of Sco X-1 had opened a new window on the universe: astronomers worldwide rushed to look through it. The AS&E researchers were quickly rejoined in the field by Friedman's group at NRL, who, in April, 1963, not only confirmed the existence of the strong source in Scorpius but also reduced the uncertainty in its position to about one degree (fig. 1.1). During the same sounding rocket flight, the NRL group found evidence for a second source, about one-eighth as strong as Sco X-1,

Fig. 1.1. The location of Sco X-1. (a) Discovery measurements (Giacconi *et al.*, 1962). Geiger counter count rate plotted versus azimuth angle. The peak in both counters at 195° represents the detection of the first extra-solar X-ray source. (*Courtesy R. Giacconi.*) (b) Confirmation measurements (Bowyer *et al.*, 1964a). The tracks of eight scans across the Scorpius region are indicated by the straight lines, labelled at intervals with count rates per 0.09 seconds. The dashed circles indicated equal intensity contours and the cross shows the most probable source position. (*Courtesy S. Bowyer. Reprinted by permission from Nature, vol. 201, p. 1307. © 1964 Macmillan Magazines Ltd.*) (c) Modulation collimator positions (Gursky *et al.*, 1966) (inset figure) compared with those of earlier experimenters. Of the four possible collimator positions, the central pair were most probable. The optical counterpart of Sco X-1 (Sandage et al., 1966) lies at RA = 16h 17m 04s, Dec = −15° 31′ 15″. (*Courtesy H. Gursky.*)

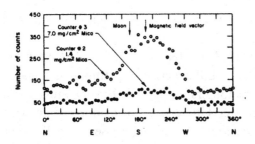

(a)

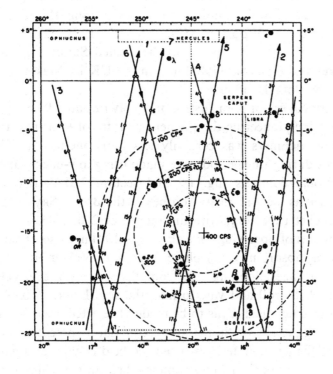

(b)

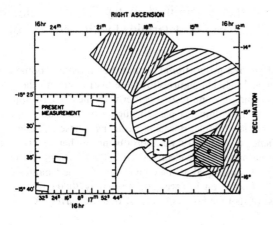

(c)

in the general direction of the Crab Nebula, the supernova remnant (SNR) of 1054 AD (Bowyer *et al.*, 1964a).

By 1967, the US groups at AS&E, MIT, NRL, the Lawrence Livermore Laboratories and Lockheed Research Labs were competing in the search for cosmic X-ray sources with astronomers in the UK, Europe and Japan. The Space Physics Group at Leicester University, for example, had been active in solar X-ray studies since the late 1950s (Russell and Pounds, 1966). Turning their attention to the wider sky, the Leicester group undertook galactic surveys in the southern hemisphere, using the Anglo–Australian rocket range at Woomera (Cooke *et al.*, 1967). The early institutional history of X-ray astronomy in the United States has been described by Tucker and Giacconi (1985) and in the UK by Massey and Robins (1986) and by Pounds (1986).

The first observational era of X-ray astronomy may be dated from 1962 to 1970. These were the pioneering days, when the ratio of known X-ray sources to X-ray astronomers was certainly less than one-to-one. The principal discoveries during this period were made using low-background large-area proportional counters (fig. 1.2 and section 2.3) carried on sounding rockets such as the American Aerobee or the British Skylark. A typical sounding rocket trajectory, with an apogee of 200–250 km, provided some five minutes of observing time above the 100 km or so of 'soft' (i.e. low-energy) X-ray absorbing atmosphere. That is, $t = 300$ s in eq. (1.1). The experimental packages, looking out from the side of the vehicle, initially used the rocket's roll and yaw to map the sky. Later, stabilised rockets allowed slow, controlled scans to be made of selected areas of the sky. A mechanical 'egg-crate' or 'honeycomb' collimator, preceding the entrance window of the gas counter, restricted the field-of-view and gave these non-imaging ($A_s = A_b$ in eq. 1.1) instruments their directional sensitivity. Mechanical collimator design is outlined in fig. 1.3. The angular location of a point source of flux F can be determined with a collimator of

Fig. 1.2. The Leicester large-area proportional counter flown from Woomera on Skylark SL 723, 12 June, 1968. The figure shows the instrument (in the triangular nosecone) undergoing ground tests together with the Skylark upper body. X-rays transmitted by the collimator pass through a thin plastic window and are absorbed within the gas volume, releasing electrons. These electrons accelerate under the influence of an electric field, producing further electrons by collisional ionisation, until a measureable electrical pulse (proportional in magnitude to the X-ray energy) is developed at the counter anode (see Chapter 2). The detector area ($A_s = A_b$ in equation 1.1) was approximately 3000 cm², making this instrument the largest of its type flown up to that time. (*Courtesy K. A. Pounds.*)

Fig. 1.3. Mechanical collimator design. (a) Plan view of regular square-section mechanical collimator. Open area fraction is given by $A_0 = (D/D+d)^2$. Open areas must generally be restricted to 75–80% to preserve rigidity of the array. Collimators with circular, hexagonal and rectangular ('slit') apertures have been used in X-ray astronomy: their relative merits are discussed by Giacconi, Gursky and Van Speybroeck (1968). Collimator walls, thickness d, are usually formed from beryllium, aluminium or stainless steel. (b) Vertical section across line AB. Soft (~ 1 keV) X-rays incident at angles θ greater than θ_c (where $\theta_c = \text{arccot}\,(L/D)$) are absorbed in the collimator walls. Very low energy X-rays may be transmitted through the collimator by multiple reflection. At energies above about 20 keV, thick walls are required to completely attenuate the flux and prevent the collimator becoming transparent at all angles of incidence. (c) Point source transmission function $T(\theta)$ for the collimator of (a) and (b). For small θ: $T(\theta) = A_0\,(1 - \theta/\theta_c)$. The fwhm transmission for this collimator, $\theta_{\frac{1}{2}}$, (determining the angular resolution (equation 1.2) and field of view of the collimated X-ray detector) equals θ_c. For practical large-area collimators, $L/D \lesssim 60$ and $\theta_c \gtrsim 1°$. The response of such a collimator to extended sources is discussed by Giacconi, Gursky and Van Speybroeck (1968).

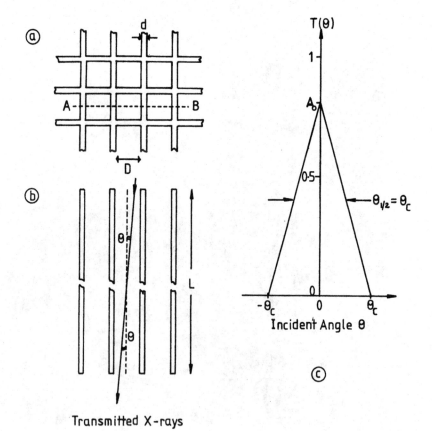

Transmitted X-rays

full-width-at-half-maximum (fwhm) transmission in the scan direction $\theta_{\frac{1}{2}}$ to an accuracy $\delta\theta$ where (Peterson, 1975):

$$\delta\theta = \theta_{\frac{1}{2}} \, (F_{min}/F) \tag{1.2}$$

The Leicester instrument described by Cooke *et al.* (1967), for example, had a rectangular aperture, geometrically collimated to $30° \times 30°$. The NRL detector described by Bowyer (1965) had a circular aperture with $10°$ fwhm transmission. Equation (1.1) tells us that the best development strategy for such instruments is to make them as large as possible (maximise A_s). This route was vigorously followed. The total effective area (product of Q and A_s) for the pioneering AS&E Geiger counters was $\sim 20 \, cm^2$: within a few years, gas detectors a hundred times larger had been constructed and flown (fig. 1.2).

High-altitude balloons provided an alternative to sounding rockets for observations of the 'harder' (i.e. higher-energy) component of the cosmic X-ray fluxes. Balloon X-ray astronomy is conducted at altitudes above 40 km, where the residual atmosphere has an effective column density of only $3 \, g \, cm^2$ (Peterson *et al.*, 1972) and is thus transparent to X-ray energies $E > 20$ keV. In a typical early balloon flight, Bleeker *et al.* (1967) used a sodium iodide scintillator crystal to measure the 20–130 keV energy spectrum of sources in the constellation of Cygnus. All scintillators (Chapters 2 and 5) rely on the production by the incident X-rays of a 'flash' of visible light which may be registered using a photomultiplier of some kind. The observing time for this early experiment was forty minutes ($t = 2400$ s in eq. 1.1). Although later balloon flights provided many hours of continuous source coverage, such experiments were, and are, restricted to an energy régime where source fluxes fall steeply with increasing X-ray energy. In the case of the Crab Nebula, $dF/dE \propto E^{-2.05}$ for energies between 10 and 100 keV. The impact of balloon experiments with scintillators (section 5.2) and cooled germanium detectors (section 4.4) was, therefore, by no means in proportion to their longer observing times.

By the end of the sounding rocket era, some 30–40 discrete sources were known. A 1968 review (Giacconi *et al.*, 1968) was forced to admit that there were: '... very few precise measurements in X-ray astronomy, a situation to be expected in a field in which each major experimental group is limited to about ten minutes of observation a year'.

Of the known sources, only a few could be confidently identified with any optical or radio counterpart. The first such identification was made by the remarkable method of lunar occultation in July, 1964. The NRL group (Bowyer *et al.*, 1964b) observed the Crab Nebula during its nine-yearly eclipse by the limb of the moon. From the gradual decrease in observed count rate, the NRL researchers were able to identify the X-ray source they had observed the year before with the optically and radio-bright Crab SNR.

Given the publicised aim of the original AS&E flight of June, 1962, it is somewhat ironic that the first precise positional measurement in X-ray astronomy used the moon as a dark occulting disc, moving against the X-ray-bright sky.

Despite the very high (>0.5 arcsecond) angular resolution of the lunar occultation technique, eclipses of bright sources occurred too infrequently for it to make a major impact on the source identification problem. The limited duration of sounding rocket flights dictated that target positions already had to be known rather accurately – more accurately than could be achieved with the standard scanning proportional counter experiments (eq. 1.2 and fig. 1.3). A number of bright galactic sources *were* observed during lunar eclipses in the early 1970s: the source GX3 + 1 (the numbers refer to its galactic latitude and longitude in degrees), observed on two separate sounding rocket flights by the Leicester group and by the Mullard Space Science Laboratory (MSSL) of University College, London, remains the most precisely X-ray-located source of all (Janes *et al.*, 1972). Lunar occultation measurements from an orbiting satellite were made for the first (and, to date, last) time in 1972: the source was GX5-1 (Janes *et al.*, 1973). Although the European EXOSAT Observatory (which evolved from a 1960s' study called HELOS – the 'LO' standing for lunar occultation) was launched into a highly eccentric, near polar orbit specially chosen to facilitate occultations over a large fraction of the celestial sphere, no such measurements were made during the satellite's lifetime.

The milestone identification of Sco X-1 with a faint blue variable star (Gursky *et al.*, 1966; Sandage *et al.*, 1966; fig. 1.1) resulted, not from the use of lunar occultation, but from a novel refinement of the collimator technique.

Equation (1.2) tells us that, in order to improve the angular resolution of a mechanically collimated instrument, all we have to do is reduce $\theta_{\frac{1}{2}}$, the fwhm transmission of the collimator. Resolution of at least an arcminute is needed to confidently identify X-ray and optical or radio sources by positional coincidence alone. Unfortunately, construction of large-area collimators for the arcminute régime poses severe engineering problems, chiefly with regard to the parallel alignment of all the collimator channels. Reduction of $\theta_{\frac{1}{2}}$ in a scanning instrument also reduces the point source observing time per scan, and hence the instrumental sensitivity. This arises from the fact that the fwhm transmission essentially determines the field-of-view.

The modulation collimator (MC), first described as an instrument for X-ray astronomy by Oda (1965), achieves sub-arcminute angular resolution while retaining a large field-of-view. An arrangement of planes of wires (fig. 1.4) is aligned perpendicular to the viewing axis of a proportional counter detector. The transmission function of such a wire collimator consists of a series of narrow bands (Bradt *et al.*, 1968; Giacconi *et al.*, 1968) so that when the instrument is scanned across the region of interest a multiplicity of possible source positions results. In the AS&E/MIT experiment described by Gursky *et al.* (1966) this problem was overcome by using two four-grid collimators with slightly differing periods in a 'vernier' arrangement. In general, a unique MC source position determination depends on the availability of data from other instruments.

In the late 1960s and early 1970s, very considerable efforts were directed into the optimisation of modulation collimators. The rotation modulation collimator (RMC) (Schnopper and Thompson, 1968) is a two-grid collimator which, centred on the rotation axis of a spinning vehicle, modulates the flux from a point source in such a way that (i) the frequency of the

Fig. 1.4. Schematic representation of a four-grid modulation collimator. The three outer grids (extending into the diagram) are separated from the inner grid by distances L, $L/2$ and $L/4$. For an n-grid collimator, separations are found from $L/2^j$, where $j=0, 1, ..., n\text{-}2$. The transmission function, in the limit of small off-axis angles ψ consists of a triangular response of fwhm s/L repeating with a period $2^{n\text{-}2}$ (s/L) (Giacconi, Gursky and Van Speybroeck, 1968). For the AS&E/MIT Sco X-1 experiment (fig. 1.1) $d=2s=0.25\,\text{mm}$ and $L=61\,\text{cm}$, producing a transmission band of $40''$ fwhm.

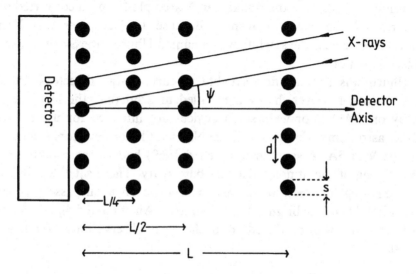

modulation depends on the off-axis distance of the source (r), and (ii) the phase of the modulation depends on the remaining polar coordinate (θ). RMCs have the advantage of being unaffected by source confusion, but reduction of the data by Fourier analysis relies on the source flux remaining constant during the observation.

The variable-spacing modulation collimator, developed at Leicester (Adams *et al.*, 1972), is a device with one fixed wire grid and a second grid of the same pitch motor-driven in the direction perpendicular to the grid planes. With the instrument mounted on a stabilised star-pointing sounding rocket, this variation of the grid spacing (L in fig. 1.4) allows images to be reconstructed rather more simply than in the case of the RMC.

In all MC designs, however, accuracy in the measurement of source position is achieved at the cost of decreased sensitivity. In a scanning n-grid device, for example, the observing time per resolution element is $1/2^{n-2}$ times that of a conventional large-area proportional counter (Peterson, 1975). It is instructive to note that no second flight of the AS&E/MIT four-grid collimator was ever made. In the limited observing times available with sounding rockets, there were simply no other sources bright enough to yield a useful signal-to-noise ratio.

Only with the coming of the first X-ray satellites did the ingenious instrumentation developed during the 1960s begin to reap the full harvest of the X-ray sky.

1.3   The small satellite era

The first satellite dedicated to X-ray astronomy, NASA's Small Astronomy Satellite (SAS) A, was launched into equatorial orbit on 12 December, 1970, from the Italian San Marco platform – a converted oil rig moored off the coast of Kenya. Because 12 December was Kenyan independence day, the satellite was renamed Uhuru – the Swahili word for freedom – in orbit.

Uhuru was not the first satellite to carry X-ray detectors. Previous instruments, however, had either failed in orbit (e.g. the large detector array on OAO-1) or were small devices not intended for use in cosmic X-ray astronomy. For example, the NaI scintillator detectors (section 5.2) on the Vela 5A, B satellites (launched 1969) had as their main aim the monitoring of the atmospheric test ban treaty. These satellites, by virtue of their long lifetimes, were able to produce an unsurpassed record of the variability of the brightest X-ray sources. An animated film – *The X-ray sky 1969–76* – was produced from the Vela observations (Terell *et al.*, 1984).

Quite simply, Uhuru revolutionised X-ray astronomy, not only because of the longer observing times available from orbit (see eq. 1.1) but also because source variability could be examined on all timescales from hours to months. The principal aim of the mission (Tucker and Giacconi, 1985) was to produce the first all-sky X-ray survey: in this, it was abundantly successful. The fourth Uhuru catalogue, the final compilation of data from the satellite's two scanning proportional counter detectors ($0.5° \times 5°$ and $5° \times 5°$ fwhm collimation; Jagoda *et al.*, 1972) contained some 339 2–6 keV source positions (4U designation; Forman *et al.*, 1978). The faintest of these was some 10 000 times fainter than Sco X-1. Identifiable classes of X-ray source – binary stellar systems, for example – appeared for the first time in the Uhuru data.

Uhuru was followed by a number of small X-ray astronomy payloads, notably: Copernicus (the third Orbiting Astronomical Observatory, launched August, 1971); the Dutch ANS (Netherlands Astronomy Satellite, launched August, 1974; Brinkman *et al.*, 1974); the British Ariel V (October, 1974; Smith and Courtier, 1976); and Uhuru's direct successor, SAS-C. A complete list of early missions is given by Peterson (1975).

Ariel V, illustrated in fig. 1.5, carried six separate experiments, four pointing along the satellite's spin axis and two scanning the sky from the side of the spacecraft.

Experiment A, provided by MSSL and the University of Birmingham, was a rotation modulation collimator consisting of two stainless steel grids. This experiment had a $17°$ fwhm field-of-view and an ultimate angular resolution of $\sim 1'$. Experiment F, built by Imperial College, London, was a scintillation counter (Carpenter *et al.*, 1976) sensitive to hard (26– 1200 keV) X-rays. The central element of this instrument was a large (8 cm^2 area) CsI(Na) crystal 'actively collimated' (section 5.2.2.1) in order to reduce background. Experiment G, the all-sky monitor (ASM) built by NASA's Goddard Space Flight Center (GSFC) (Desai and Holt, 1972; Holt, 1976) consisted of a pair of pinhole cameras with $4°$ fwhm fields-of-view, looking out in opposition along the satellite's spin axis (fig. 1.5). These small counters (0.6 cm^2 aperture) mapped out some 80% of the sky per 100 minute orbit. The ASM was used to keep watch for X-ray transients – sources which suddenly and dramatically increased in brightness. A pinhole camera is the simplest in the family of coded aperture or shadow-mask cameras, described in more detail in section 1.5.

MSSL's Experiment C (Sanford and Ives, 1976) had as its aim the detailed study of source spectra. The crucial figure of merit for any X-ray spectrometer is its energy resolution $\Delta E/E$. Suppose the spectrometer is illuminated by a monochromatic beam of X-rays, energy E. The energy which can be inferred from the spectrometer output will not be unique, but will follow some distribution with a most probable value E and full-width-at-half-maximum value ΔE. Energy resolution (in the case of solid state detectors (Chapter 4), the unnormalised width ΔE is often referred to as the energy resolution) is therefore a measure of the instrumental blurring of source spectral detail.

Experiment C was a xenon–methane proportional counter (3.5° fwhm field-of-view; 100 cm² geometric area) sealed by a thick beryllium window.

Fig. 1.5. The Ariel V satellite (Smith and Courtier, 1976). Launched in 1974 from the San Marco launch complex, Ariel V re-entered in March, 1980, after 30 152 orbits. Some 1.5 m tall, it weighed only 130 kg. The satellite's payload is described in the text. (*Courtesy G. M. Courtier.*)

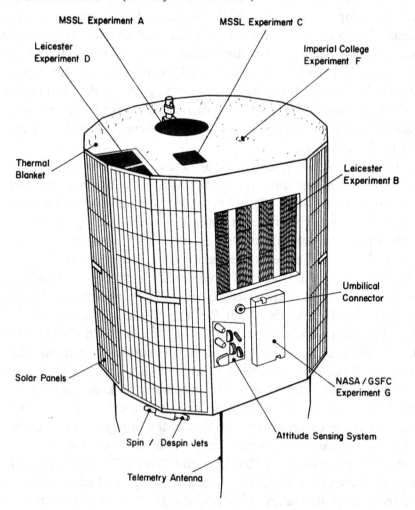

Its ability to point at sources for days at a time gave it high sensitivity. Its energy resolution, $\Delta E/E$, however, was unfortunately typical of the gas detectors which were the 'workhorses' of X-ray astronomy's first decade:

$$\Delta E/E = 0.4/E^{\frac{1}{2}} \text{ fwhm (28\% at 2 keV; 16\% at 6 keV)}$$

Unfortunately, that is – because although the early excitement of X-ray astronomy lay in the location (and timing) of new sources, it was widely recognised that much of the astrophysics would lie in the source spectra. Proportional counter energy resolution was simply not good enough for the detailed study of narrow line emission from highly ionised oxygen, silicon, iron and the other astrophysically abundant elements.

Ariel V therefore carried a second type of X-ray spectrometer, capable of much better energy resolution – a flat crystal spectrometer (fig. 1.6a). A similar instrument was flown on ANS. Crystal spectrometers use the law enunciated by W. L. Bragg in 1912 to describe the reflection of X-rays from regularly spaced atomic planes:

$$n\lambda = 2d \sin \theta \tag{1.3}$$

Here, n is the order of the reflected beam, λ is the X-ray wavelength, d is the spacing between crystal lattice planes and θ is the grazing angle of incidence to the crystal surface.

Crystal spectrometers had previously been flown on sounding rockets to study the sun and the brightest cosmic sources. Pounds (1971), for example, reports an (unsuccessful) attempt to detect the 6.7 keV line emission from helium-like iron in the spectrum of Sco X-1, using a large-area lithium fluoride crystal. The first successful high-resolution line detection (O VII Lyman α in the spectrum of the SNR Puppis A) from a non-solar source was in fact made by the MSSL group in October, 1974 (Zarnecki and Culhane, 1977).

The Ariel V crystal spectrometer (Experiment D: Griffiths *et al.*, 1976) contained two large (234 cm²) composite crystal panels, one of graphite ($2d = 6.708$ Å), the other of lithium fluoride ($2d = 4.026$ Å). These were optimised for the detection of silicon and sulphur line emission around 2 keV (graphite) and for iron emission in the 6–7 keV band (LiF). They were capable of energy resolutions

$$\Delta E/E = 0.3\% \text{ (at 2 keV); } 0.7\% \text{ (at 6 keV)}$$

much better than the corresponding figures for any gas counter.

With Bragg angles close to 45°, the instrument could also be used as a polarimeter (Gowen *et al.*, 1977). At such angles (akin to the Brewster angle for optical wavelengths) only X-rays with their electric vectors perpendicular to the plane of incidence (s-polarised) can undergo Bragg reflection. Rotation of the crystal relative to the line of sight then produces a modulation of the count rate from which the degree of linear polarisation

Fig. 1.6. Flat crystal, curved crystal spectometers. (a) The Ariel V flat crystal spectrometer–polarimeter (Griffiths *et al.*, 1976) and a schematic of its operation. If a collimated beam of X-rays containing a range of wavelengths λ impinges on a flat crystal, only that λ_1 for which the angle of incidence θ_1 satisfies the Bragg law (equation 1.3) is reflected into the detector. The detector area must equal the area of the collimated beam, A_s. The crystal must be rotated to a new Bragg angle θ_2 to examine wavelength λ_2. (*Courtesy K. A. Pounds and the Royal Astronomical Society.*) (b) Exploded view of the OSO-8 curved crystal polarimeter (Novick *et al.*, 1977) and a schematic of its operation. The composite crystals are mounted on sectors of parabolic surfaces. Predominantly s-polarised X-rays incident at Bragg angles of 40–50°, and consequently with a small spread in energy, are focused to a small spot. The detector area (the area for intrinsic background) can therefore be much smaller than A_s. (*Courtesy R. Novick.*)

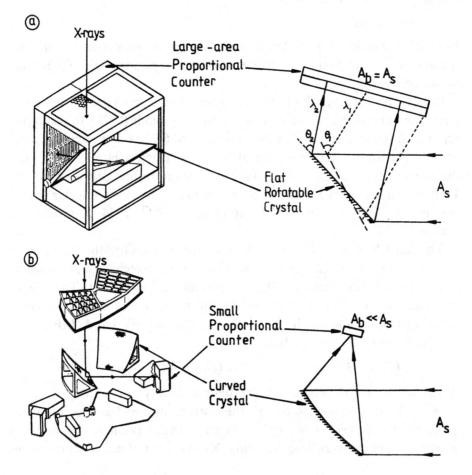

of the source can be derived. The second 'classical' method of detecting linear X-ray polarisation is to observe the anisotropic Thomson scattering from a lithium, lithium hydride or beryllium block (Landecker, 1972; Lemen *et al.*, 1982). Both techniques were used in the first positive detection of X-ray polarisation, from the Crab Nebula, in 1971 (Novick *et al.*, 1972). An important figure of merit in all polarimeters is the modulation contrast *M*:

$$M = (N_{max} - N_{min})/(N_{max} + N_{min})$$

Here N_{max} and N_{min} are the maximum and minimum count rates observed during rotation about the line of sight to a 100% linearly polarised source. For Bragg polarimeters, $M = 0.96$–0.99. For Thomson polarimeters, $M \sim 0.25$.

Results from Experiment D, however, were generally disappointing, even for long observations on the brightest sources. The sensitivity of any Bragg spectrometer can be properly estimated from the crystal's integrated reflectivity $\Delta\theta$ (Novick *et al.*, 1977), a quantity which contains information on both X-ray reflectivity and bandwidth. For Experiment D, $\Delta\theta = 2$–9×10^{-4} radians, depending on crystal type and angle setting. The problem of low throughput, common to all crystal spectrometers, was compounded in the case of Experiment D by the use of flat crystals. Use of a flat crystal demands the use of a large detector (fig. 1.6a). In terms of eq. (1.1), the collecting areas for source photons (A_s) and for detector intrinsic background (A_b) are then the same.

The Columbia University polarimeter, flown on the OSO-8 satellite in 1975 (Novick *et al.*, 1977), using graphite panels of a similar size to those on Ariel V, with similar integrated reflectivities, proved a much more sensitive instrument because its crystals were curved. As shown in fig. 1.6b, the resulting Bragg focusing effect allowed the detector (background collecting) area to be much reduced. Inspection of eq. (1.1) shows that crystal curvature yields a sensitivity advantage $(A_b/A_s)^{\frac{1}{2}}$ (in the limit of intrinsic background dominance) over a flat crystal system with the same aperture. Whereas Experiment D in its polarimetric mode could only place broad upper limits on the linear polarisation of Sco X-1 (Gowen *et al.*, 1977), OSO-8's polarimeter enabled definite estimates to be made of the degree of 2.6 keV linear polarisation for a number of sources, down to 3% for the black hole candidate Cygnus X-1 (Novick *et al.*, 1977). Polarimetric measurements provide an important means of distinguishing between source emission mechanisms.

Ariel V's remaining experiment, the Leicester Sky Survey Instrument (SSI) (Experiment B; Villa *et al.*, 1976) provides an excellent example of a non-imaging survey detector, with all the strengths and weaknesses of the breed. Like the Uhuru instruments, the SSI was a scanning, slat-collimated proportional counter ($A_s = A_b = 280\,\text{cm}^2$) with, for most of its lifetime, a 2.4–19.8 keV, $0.75° \times 10.6°$ fwhm field-of-view. As its name suggests, the purpose of the instrument was to make a complete sky survey, in the manner of Uhuru. As Ariel V spun about its axis, the side-mounted SSI (fig. 1.5) swept out a great circle on the sky, viewing a strip some 20° wide. In survey mode, which provided most of the eventual source catalogue data (McHardy *et al.*, 1982; Warwick *et al.*, 1982) counts were integrated for one orbit, the 360° scan path being divided into 1024 spatial elements (fig. 1.7a). For a one-orbit observation, *t* in eq. (1.1) corresponded to the time per orbit during which the source was within the field-of-view. Since the spin-axis position of the satellite was typically held constant for several days, data from successive orbits could be superimposed – up to 150 in some cases (Warwick *et al.*, 1982). From eq. (1.1), summation of *n* independent scans improved the sensitivity as $n^{\frac{1}{2}}$.

Each acceptable source peak in the summed-orbit data, when mapped on to the sky, constituted a 'line of position' of length $\sim 20°$ and of angular width determined by the significance of the peak above the fitted background. The intersection of lines of position at widely separated scan directions determined each source position, as shown in fig. 1.7b. The resulting error boxes (enscribing the 90% confidence contour around the most probable intersection point) ranged in area from ~ 0.01 square degrees for the strongest sources to several tenths of a square degree for the weakest; as for Uhuru, $F_{\text{min}} = 10^{-4}$ times the intensity of Sco X-1. In crowded regions of the sky, especially at low galactic latitudes (Warwick *et al.*, 1982), source confusion, alluded to in section 1.1, posed a severe problem in the data analysis. The precise position – or indeed the existence – of a weak source in the neighbourhood of a strong source could be affected by the (partly subjective) assignment of a small number of lines of position to one or the other.

In summary, the SSI, and by extension the whole class of collimated, scanning instruments, could be characterised by: excellent sky coverage,

Fig. 1.7. The Leicester Sky Survey Instrument. (a) Data from a single scan of the SSI (in fact, from Ariel V's last orbit). The peaks correspond to the passage of the named X-ray sources through the field-of-view of the detector. (b) SSI lines of position for the SNR Puppis A (4U0821-422), displayed in galactic coordinates. Intersection of these individual detections determined the source location. Circles represent the positions of other catalogued sources.

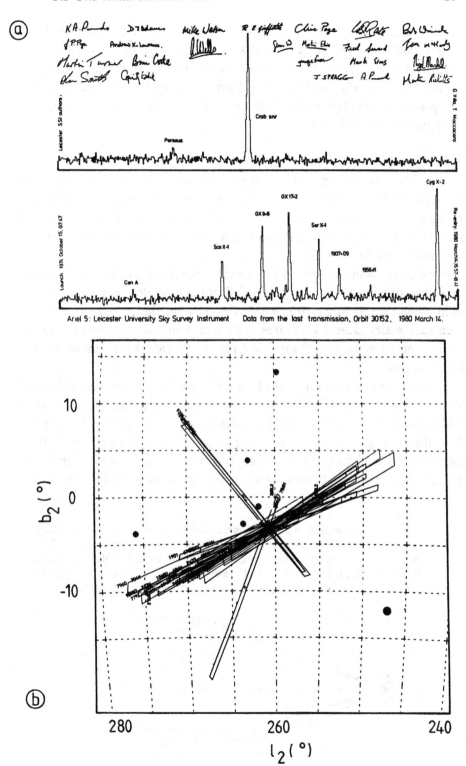

Ariel 5: Leicester University Sky Survey Instrument   Data from the last transmission, Orbit 30152, 1980 March 14.

modest angular resolution, moderate sensitivity, and a susceptibility to source confusion on angular scales of a few degrees. This last factor complicated the already tortuous process of producing sky maps. Contemporary with the SSI, on board the Copernicus and ANS satellites, were mapping instruments of a very different kind: the precursors of the first true imaging X-ray telescopes.

1.4   Imaging

The development of collimated instruments has been paralleled throughout the short history of X-ray astronomy by research into focusing systems, whose evolution we shall now sketch. Indeed, the first proposals for their use in X-ray astronomy predate the discovery of Sco X-1 (Giacconi and Rossi, 1960).

Focusing systems for soft X-ray energies are based on the total external reflection of photons incident at small grazing angles of incidence – less than some critical angle (θ_c), which depends on the composition of the reflecting material and decreases with increasing X-ray energy (Henke, 1972). For example, $\theta_c = 1°$ for 3 keV photons incident on a metal such as nickel or gold.

A paraboloid of rotation is the simplest focusing element (Giacconi and Rossi, 1960, and references therein). The image of a point X-ray source at infinity is a comatic circle in the focal plane, whose radius is proportional to the off-axis angle of the source (fig. 1.8). Such a paraboloid cannot, therefore, be used to form an image; nor can any single-reflection X-ray optic. It can, however, be used as a flux concentrator – just like the curved

Fig. 1.8. Cross-section of a paraboloidal mirror. The lower part of the figure shows focusing of X-rays, incident parallel to the axis XX', to a point F. The grazing angle of incidence to the section OO' is θ. The half-angle of the focused X-ray cone is 2θ. The upper part of the figure shows the 'nesting' of paraboloidal sections, with the same focus, in order to increase collecting area.

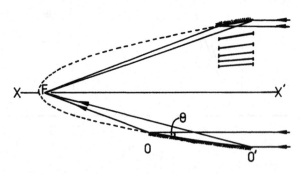

Bragg crystal considered in the previous section – allowing the size of the associated photon detector to be minimised with a sensitivity advantage $(A_b/A_s)^{\frac{1}{2}}$. The first such flux concentrators – or 'light buckets' – were flown on the Copernicus satellite (Bowles *et al.*, 1974) in 1971. The Copernicus X-ray instrument had been proposed some eight years earlier, by Boyd and Willmore at University College, London (later MSSL), working independently of Giacconi and his colleagues in the US. Two of the three Copernicus paraboloids were used in conjunction with small-volume proportional counters; at the exit aperture of the third was a *channel electron multiplier* (see Chapter 3). In aggregate, the detectors covered the energy range 0.15–4 keV with a collecting area (A_s) of 41 cm². The angular resolution was $\sim 1'$, achieved by aperture stopping of the telescopes. ANS carried a slightly larger paraboloid system (~ 70 cm² area; den Boggende and Lafleur, 1975) working in the 0.15–0.28 keV band. The area of this experiment's thin-window proportional counter was only 7 cm², giving a 'focusing advantage', $(A_s/A_b)^{\frac{1}{2}}$, of about three. Further paraboloidal mirrors were flown on the British Ariel VI satellite in 1979.

A second form of focusing collector is an array of coaligned (nested) reflecting sheets bent into parabolic curves (fig. 1.9a). Such an array has the property that the image of a point source is a line, hence the common name given to this mirror design – a 'one-dimensional' (1-d) imaging system. An early form of this collector was described by Fisher and Meyerott (1966). A later sounding rocket version was used by Gorenstein *et al.* (1971a, b), to study the X-ray structure of the Cygnus Loop SNR, with a sensitivity further increased by the introduction, probably for the first time in X-ray astronomy, of a position-sensitive detector. A four-anode gas proportional counter, with the anode wires parallel to the direction of focusing, provided a detector with four resolution elements (fig. 1.9a). Positive source detection then required that the source be 'seen' above the background level in only one of these, rather than above the background of the whole detector. In terms of eq. (1.1), A_b is equal to the area of one resolution element in any imaging system which combines focusing optics and a position-sensitive detector. Gorenstein *et al.* (1971a) compared their instrument, with an effective area for 0.15–0.28 keV X-rays of 160 cm², to a typical Uhuru- or SSI-type instrument of similar bandwidth and area (A_s) equal to 1200 cm². The focusing instrument was found to be a factor three more sensitive. An added, indirect advantage of the focusing system was that thinner polypropylene windows could be reliably used on the smaller proportional counters (increasing Q in eq. 1.1).

The first two-dimensional imaging X-ray telescope used in cosmic X-ray astronomy was described by Gorenstein *et al.* (1975). Its mirror was of the type first proposed by Kirkpatrick and Baez (1948) and was based on the principle that two successive X-ray reflections from orthogonal curved surfaces could be used to form images (fig. 1.9b). In essence, the 2-d mirror was a pair of 1-d focusing devices back-to-back, with the second rotated 90° with respect to the first. In contrast to the 1-d system, whose wide field-of-view ($\sim 10°$) suited it to scanning large areas of sky, the 2-d device was

Fig. 1.9. Kirkpatrick–Baez grazing incidence optics. (a) Schematic view of 1-d focusing collector plus detector. The line image of a point source falls on one of the four anodes in the multi-wire proportional counter (MWPC). The mirror parabolae are formed from chromium-coated bent glass plates (Gorenstein *et al.*, 1971a). (b) Small-angle reflection from two orthogonal curved surfaces. If the surfaces are parabolae of translation, a parallel beam of incident radiation (*I*) is focused to a point (*O*). A practical mirror consists of several parallel (nested) surfaces such as these (Gorenstein *et al.*, 1975, after Kirkpatrick and Baez, 1948). © *1975 IEEE* (c) X-ray image of a $2 \times 2°$ field containing the star Algol. A 25 s exposure obtained on 6 December, 1975, with a Kirkpatrick–Baez mirror and MWPC detector system of 150 cm² effective collecting area at 1 keV. Pixel size $4' \times 4'$. Image scale indicated by 16' tick. (Harnden *et al.*, 1977.) (d) X-ray image of the Virgo cluster, centred on M87. A 120 s exposure containing 1100 counts. (Gorenstein *et al.*, 1977.) (*Figures courtesy P. Gorenstein, F. R. Harnden.*)

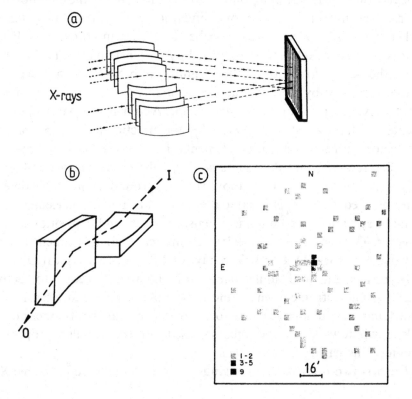

Figure 1.9. (Continued)

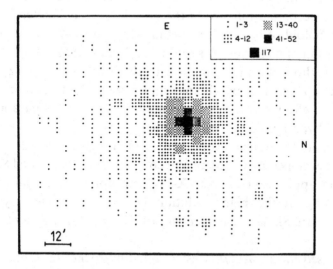

designed for the study of extended objects (clusters of galaxies) with moderate angular resolution (4′ fwhm) over a small (40′ radius) field-of-view. Focused images were detected by a multi-wire proportional counter (MWPC) (Chapter 2), position sensitive in two dimensions. The power of a 2-d imaging telescope of even modest resolution is illustrated in figs. 1.9c and d. The first of these figures shows the 0.15–1.5 keV image of Algol (β Persei) obtained on one flight of the Harvard–Smithsonian Center for Astrophysics' telescope (Harnden *et al.*, 1977). Only nine photons were detected from the resolution element coincident with Algol in a 25 s exposure. The integrated detector background per resolution element (A_b) was only 0.13 counts, however, so that the probability that this peak was a random (Poissonian) fluctuation in the background level was of order 10^{-13}.

The 'imaging advantage' $(A_s/A_b)^{\frac{1}{2}}$ for this Kirkpatrick–Baez telescope was about 30. It was flown twice on sounding rockets to make observations of the Virgo and Perseus clusters of galaxies (Gorenstein *et al.*, 1977, 1978) which were known from Uhuru days to be extended sources of X-ray emission. Fig. 1.9d is the 0.15–1.5 keV image obtained of the Virgo Cluster, which revealed the giant galaxy M87 embedded in diffuse emission – allowing a previously postulated source model (that the emission was due to a number of unresolved point sources) to be discarded.

The first images of extended galactic X-ray sources were obtained from sounding rocket flights of an imaging telescope developed jointly by the Center for Space Research at MIT and the University of Leicester X-ray Astronomy Group. The focusing optics consisted of a nested (coaxial) pair of paraboloid–hyperboloid mirrors of the kind known as Wolter Type 1 (the simple paraboloid of fig. 1.8 is sometimes known as Wolter Type 0). Such a combination of conic surfaces (fig. 1.10; Wolter, 1952a, b) provides an alternative method of imaging X-rays to the Kirkpatrick–Baez technique discussed above. The paraboloid and hyperboloid are confocal. Paraxial rays are reflected towards the common focus by the paraboloid and then by the hyperboloid towards its other focus.

Wolter's original theoretical work had been directed towards X-ray microscopy; as developed by Giacconi and co-workers in the United States it quickly found application in solar X-ray astronomy (Giacconi *et al.*, 1965), culminating in the high-resolution (2″ fwhm) Wolter Type 1 instruments on board Skylab (Vaiana *et al.*, 1973) which used X-ray film in the focal plane.

The mirrors used in the MIT/Leicester sounding rocket telescope were somewhat more modest than the Skylab solar telescopes. A thin-window MWPC registered the photon positions in the focal plane (Rappaport *et al.*, 1979). The focal length of the nickel-coated aluminium mirrors (z_0 in fig. 1.10) was 1.14 m, leading to a focal plane scale of 3′/mm. For its first

Fig. 1.10. Cross-section of Wolter Type 1 grazing incidence mirror. X-rays incident parallel to the axis XX' first strike the paraboloidal section P and are reflected towards the common focus ($F2$) of the parabola and the hyperbola $H2$. A second reflection diverts the beam towards the remaining hyperbolic focus, $F1$. The focal length of the combination, z_0, is measured from the join of the conic sections. The half-angle of the focused X-ray cone is four times the grazing angle of incidence to the mirror sections.

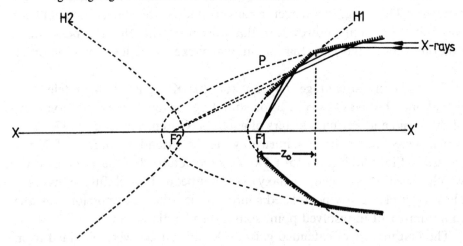

flight the angular resolution of the telescope was only 9′ (rms point source blur circle radius), determined by the contribution of the gas detector. That is, the telescope was 'detector limited' in resolution, a generally undesirable state of affairs since mirrors of high quality are generally more expensive and harder to make than detectors. Nevertheless, the telescope resolution was a factor of three better than for previous non-imaging observations of the Cygnus Loop, the large (3° diameter) SNR examined on 27 July, 1977. The image produced by two, well-separated scans across the SNR is shown in fig. 1.11. The shell structure of the remnant is clearly visible, but there is

Fig. 1.11. A 0.15–1.5 keV image of the Cygnus Loop SNR (Rappaport *et al.*, 1979). This map, containing ∼7000 counts, was obtained during a sounding rocket flight of the MIT/Leicester Wolter Type 1 telescope in July, 1977. Each cross represents 1.2 counts; the pixel size is 4′ by 4′. The map has been corrected for the effects of varying exposure and filtered to suppress statistical fluctuations. The 10′ radius circle (the telescope beam size) represents the area within which 80% of the X-rays from an on-axis point source would be imaged. (*Courtesy S. Rappaport.*)

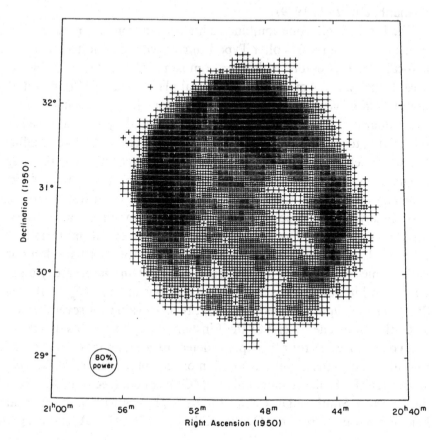

no evidence for the central point source suggested by earlier observations with 1-d imaging systems (Rappaport *et al.*, 1979, and references therein). A second flight of the MIT/Leicester telescope on 8 March, 1978, produced images of the Puppis A and IC443 SNRs with somewhat improved (4′) angular resolution (Levine *et al.*, 1979).

The sounding rocket telescopes described above were the forerunners of the first satellite-borne, 2-d imaging system – the Einstein Observatory (High Energy Astrophysics Observatory (HEAO) 2; Giacconi *et al.*, 1979), whose impact on X-ray astronomy surpassed even that of Uhuru. Immediately prior to the launch of Einstein, only a hundred or so of the known X-ray sources had precise ($<1′$) positions (Bradt *et al.*, 1979), principally obtained from the Ariel V and SAS-C RMCs and from the HEAO 1 A3 experiment (a pair of four-grid modulation collimators, each of $400\,cm^2$ effective area; Gursky *et al.*, 1978). After two-and-a-half years of Einstein operation, several thousand sources had been precisely located, encompassing the entire range of astronomical objects, from almost every class of main-sequence star (Vaiana *et al.*, 1981) to quasars of high redshift (Tananbaum *et al.*, 1979).

The Einstein telescope combined a high-resolution (2″ rms blur circle for on-axis point sources) Wolter Type 1 mirror with four nested elements and a focal plane assembly with four complementary detectors. The mirror focal length was 3.4 m, leading to a focal plane scale of 1′/mm and a $\sim 1°$ useful field-of-view. The four nested mirror shells, manufactured from fused quartz and coated with a chrome–nickel alloy for increased X-ray reflectivity, together had an effective area of $400\,cm^2$ at 0.25 keV, falling, in a manner determined by the decrease in mirror reflectivity with energy at the mean grazing angle of $\sim 1°$, to $30\,cm^2$ at 4 keV. The grazing-incidence geometry dictates that the ratio of collecting area to polished mirror area is small in any Wolter Type 1 system; for the Einstein mirrors at 0.25 keV this ratio was ~ 0.003. The combination of high-resolution (small A_b in equation 1.1), relatively large photon-collecting areas (A_s) and long observation times ($t = 10^4\,s$ per source) gave the Einstein Observatory a sensitivity several hundred times that of any previous mission ($F_{min} = 10^{-7}$ times Sco X-1). Our figure of merit, $(A_s/A_b)^{\frac{1}{2}}$, reaches values of several thousand when the lower energy end of the Einstein bandpass is considered.

Two of Einstein's four focal plane detectors were image-forming devices. Best matched to the angular resolution of the optics was the *high resolution imager* (HRI), a microchannel plate (MCP) camera (section 3.4.2) built by the Harvard–Smithsonian Center for Astrophysics with support from the Leicester X-ray Astronomy Group (Kellogg *et al.*, 1976). A channel plate is essentially an array of miniature photomultipliers. The HRI provided arcsecond imaging over the central 25′ of the telescope field-of-view, albeit with a low detection efficiency and without any intrinsic energy resolution. The second imaging detector was a high efficiency multi-wire Ar–Xe–CO$_2$ proportional counter – the *imaging proportional counter* (IPC: Harvey *et al.*, 1976; Humphrey *et al.*, 1978), developed from the sounding rocket detector of Gorenstein *et al.* (1975). The IPC provided both moderate energy resolution ($\Delta E/E = 100\%$ above 1.5 keV in energy) and 1′ imaging at the centre of a 1° diameter field-of-view (see section 2.4.2).

The remaining elements of the Einstein focal plane assembly were non-imaging spectrometers which took advantage of the mirror's power as a flux concentrator. The *focal plane crystal spectrometer* (FPCS: Canizares *et al.*, 1978), built at MIT, was a curved crystal Bragg spectrometer with six alternative diffractors. Although capable of excellent energy resolution, the various modes of the FPCS were all rather insensitive, with effective areas of order 0.5 cm^2.

Intermediate to the IPC and FPCS in terms of energy resolution, but with the high efficiency of the former detector, was the *solid state spectrometer* (S^3: Joyce *et al.*, 1978) a non-dispersive spectrometer based on the creation, by focused X-rays, of electron–hole pairs in a lithium drifted silicon (Si(Li)) crystal. The S^3 was the only part of the Einstein payload which required cryogenic cooling, down to a temperature of 100 K. Solid state X-ray detectors are discussed in Chapter 4.

Finally, the so-called *objective grating spectrometer* (OGS: the term 'objective' is, strictly speaking, inappropriate here, since the gratings were not the first elements in the optical path) could be placed in the X-ray beam behind the high-resolution mirrors in order to produce broad-band dispersed spectra on the surface of the HRI, according to the grating equation

$$m\lambda = d(\sin\varphi_i + \sin\varphi_d) \qquad (1.4)$$

Here, m is the integer order of the spectrum (± 1 for the OGS), d is the grating period, φ_i is the angle of incidence (relative to the optical axis), φ_d is the dispersion angle and λ is the X-ray wavelength. Transmission grating operation is sketched in fig. 1.12. The two OGS gratings, prepared by the Space Research Laboratory at Utrecht, had gold line densities ($1/d$) of 500 and 1000 lines/mm (Seward *et al.*, 1982). Maximum sensitivity of the HRI–OGS combination was in the soft X-ray band below 0.25 keV, where, however, the total effective area (mirror effective area times HRI efficiency times grating efficiency) was only about 1 cm^2. The OGS wavelength resolution $\lambda/\Delta\lambda$ (the reciprocal of its energy resolution) was about 50.

Transmission gratings had previously been used in solar X-ray astronomy (Vaiana *et al.*, 1968).

The scientific yield of the Einstein mission is discussed in full by Tucker and Giacconi (1985). Only one example of the observatory's quite revolutionary capabilities will be given here. Fig. 1.13 is a contour plot of the 0.2–4 keV intensity distribution across the Cassiopeia A SNR, recorded with the HRI. The radius of the inner shell is about 100″. A plateau of emission with radius 140″ is also visible (Murray *et al.*, 1979; Fabian *et al.*, 1980). The point to note is that this entire detailed image has an angular extent corresponding to only 2×2 resolution elements of the early sounding rocket maps shown in figs. 1.9 and 1.11.

1.5 Post-Einstein: the modern era

Something of the major goals of modern X-ray astronomy can be gauged from an examination of three of the most important satellite missions now under development: ROSAT, AXAF and XMM.

Successful as the Einstein mission was, it imaged, in total, only about 1% of the celestial sphere. The first imaging sky survey, two to three orders of magnitude more sensitive than those of Uhuru or Ariel V, will be carried

Fig. 1.12. Transmission grating spectrometer (Brinkman *et al.*, 1985). The grating annulus, composed of many small grating elements, is positioned in the converging X-ray beam of a Wolter Type 1 mirror. Having undergone two reflections (at *a*, *b*) X-rays are diffracted by the parallel grating rulings into first order (F_1, F_{-1}) and higher orders on the surface of an imaging detector. F_0 is the zeroth order. Typical first order grating efficiencies are $\sim 10\%$. Grating spectrometers may be regarded as devices of constant $\Delta\lambda$ yielding highest wavelength (energy) resolutions at long wavelengths (low energies). Spectrometers of this type have flown on the Einstein (Seward *et al.*, 1982) and EXOSAT (Brinkman *et al.*, 1980) observatories and are proposed for the AXAF and SPECTROSAT missions. (*Courtesy A. C. Brinkman.*)

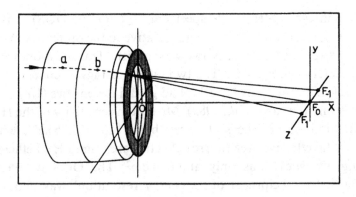

out by ROSAT (Röntgensatellit), a West German national project in which the USA and UK are collaborators (Trümper, 1984). In its survey mode, the main ROSAT X-ray telescope (XRT) will consist of a large (focal length $z_0 = 2.4$ m, geometric area $A_s = 1140$ cm^2) Wolter Type 1 nested mirror with an Ar–Xe–CO$_2$ MWPC in the focal plane (Pfeffermann and Briel, 1985: see section 2.4.4). The 6–80 Å survey is planned to last six months and will be followed by a pointed phase of the mission lasting one year or longer, in which the full angular resolution of the XRT will be exploited by the use of an upgraded copy of the Einstein HRI at the mirror focus.

The ROSAT mission, now due for launch in February, 1990, will be a landmark in a second sense: the UK *wide field camera* (WFC), coaligned

Fig. 1.13. A 0.2–4 keV X-ray image of the Cas A SNR obtained by the Einstein HRI, contoured at constant intensity. Data from the 32 500 s exposure has been filtered using the maximum entropy method (MEM) prior to contouring. The angular resolution of the image is approximately 4″ fwhm. (*Courtesy R. Willingale.*)

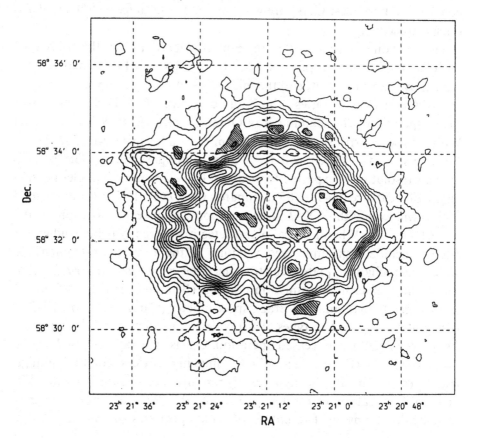

with the XRT, is expected to perform the first ever sky survey in the 60–200 Å (XUV) band. The WFC has been built by a consortium of five UK institutions (the Universities of Leicester and Birmingham, MSSL, Imperial College and the Rutherford Appleton Laboratory). It consists of a short focal length (0.53 m) Wolter–Schwarzschild Type 1 grazing-incidence mirror (Willingale, 1984), with a 5° field-of-view and an on-axis collecting area of 511 cm^2, and a large microchannel plate detector (section 3.4.3). The first source with emission predominantly in the XUV (the hot white dwarf HZ 43) was discovered as recently as 1975 (Lampton *et al.*, 1976a). For many years it had been believed that the XUV band was observationally inaccessible, due to heavy absorption by the interstellar medium: presently only a small number of such sources are known. The WFC, together with the Extreme Ultraviolet Explorer (EUVE) Satellite built by the Space Sciences Laboratory of the University of California at Berkeley (Bowyer *et al.*, 1981a), is expected to increase the number of known XUV sources to at least 1000.

A follow-up mission to ROSAT, SPECTROSAT, is already under consideration in West Germany with, as the name suggests, the emphasis on spectroscopy.

The most direct descendant of the Einstein Observatory is the US AXAF (Advanced X-ray Astrophysics Facility: Weisskopf, 1985), planned for launch by the Space Shuttle in the mid-1990s. AXAF will be one of the so-called 'Great Observatories', the X-ray counterpart of the Hubble Space Telescope and of the Gamma Ray Observatory (GRO). With six nested Wolter Type 1 mirror shells ($A_s = 1700$ cm^2) and a ten metre focal length (cf. Einstein's 3.4 m) AXAF will be a giant among satellites. The on-axis angular resolution is expected to be about 0.5″. The use of smaller grazing angles of incidence (0.45°–0.85° compared with 0.68°–1.17° for the Einstein optics) will extend the AXAF mirror response to cover the astrophysically important 6–7 keV band. Focal plane detectors currently being studied for AXAF include some improved versions of the Einstein instruments: a larger microchannel plate detector, the *high resolution camera* (section 3.4.2); more efficient transmission gratings (Brinkman *et al.*, 1985; Canizares *et al.*, 1985a); another Bragg spectrometer (Canizares *et al.*, 1985b). Also under study are two types of X-ray detector which have evolved in the few years since Einstein ceased operation. One is an array of cooled *charge coupled devices* (CCDs: section 4.6), imaging devices combining high spatial resolution with the good energy resolution of silicon. The other is a single photon calorimeter (Chapter 6), probably the most important new detector development of recent years, which promises to combine in one instrument the energy resolution of a crystal spectrometer with the high efficiency of non-dispersive detectors. Calorimeters work by sensing the temperature rise induced in a small cryogenically cooled mass by the absorption of a single X-ray photon (representing only $\sim 10^{-15}$ joules of energy!).

The scientific capabilities of AXAF are fully described in a special issue of *Astrophysics Letters and Communications* (vol. 26, no. 1, 2, 1987).

The 'high-resolution' route to higher sensitivity which AXAF represents is an ambitious one (NASA's call for focal plane detector proposals modestly stated that one of the scientific aims of AXAF was to 'understand the history and evolution of the universe'). Undoubtedly it will also be very expensive. The fabrication, mounting and calibration of large, precisely figured mirrors, which must be smooth to a roughness level of less than 15 Å, all present formidable engineering difficulties (Wyman *et al.*, 1985). Willingale (1984) and Aschenbach (1985) give comprehensive accounts of the fabrication of high-resolution grazing-incidence optics.

In Europe, an alternative, though no less challenging, route to high sensitivity is being pursued. The European Space Agency X-ray spectroscopy 'cornerstone' mission (usually known as XMM – the X-ray multi-mirror: Bleeker, 1985) compromises angular resolution (i.e. A_b) for the sake of an enormous collecting area (A_s). XMM will consist of a number (probably three) of identical mirror modules, each with the collecting area of AXAF, but with only 30″ on-axis angular resolution. XMM's focal plane instruments – possibly *gas scintillation proportional counters* (GSPC: see section 2.5), CCDs, Bragg crystals or a novel form of X-ray grating operating in reflection mode (Hettrick and Kahn, 1985) – will be devoted to spectroscopic and timing studies. XMM reflects a worldwide drive towards the production of large collecting areas from low-cost, mass-produced optics. In the United States, the concept of combining data from a number of identical, coaligned telescopes goes by the name of LAMAR (*large area modular array of reflectors*). An eight module Kirkpatrick–Baez telescope based on commercial float glass plates is under development for a short Shuttle flight (Gorenstein *et al.*, 1985). An excellent evaluation of the various mass production approaches to mirror fabrication (including replication from polished mandrels, as for the EXOSAT low-energy telescopes (de Korte *et al.*, 1981a) and approximation of the Wolter Type 1 geometry by a single cone (Serlemitsos *et al.*, 1984)) is to be found in the report of the ESA XMM telescope working group (Aschenbach *et al.*, 1987b).

A possible 'precursor' to XMM, the Joint European X-ray Telescope

(JET-X), comprising perhaps one replicated mirror module and its associated CCD detector, is currently under consideration by a consortium of
Western European groups for flight on the Soviet Spectrum X mission.
Western participation in the Soviet space astronomy programme, which
currently includes the successful Mir/Kvant high-energy observatory, is
likely to increase in the next decade (Smith, 1987).

Both AXAF and XMM, because of their expense and technical complexity, as well as because of external factors such as the *Challenger* Shuttle
disaster, have been subject to much frustrating descoping and/or delay (see
fig. 1.14).

The current observational status of X-ray astronomy can best be
summarised in terms of the five source parameters accessible to
observation: intensity, variability, direction, energy-dependence and
polarisation.

The first two of these properties have not featured prominently in the
present discussion simply because they have rarely been major instrument
'drivers'. The ability to measure source intensities is common to all
(properly calibrated) instrumentation, while 'time-tagging' photon arrival
times down to the microsecond level has been routine with all electronic
detectors from the collimated proportional counter onwards. Only CCDs
of all the modern electronic detectors are not strictly speaking 'real-time'
photon counters, in the sense that evidence of a photon's arrival does not
immediately appear at the output node (see section 4.6). Large-area
proportional counters such as the Large Area Counter, built by the
Leicester group and the Rutherford Appleton Laboratory for the Japanese
Ginga mission (see section 2.3), and similar instruments planned for the US
X-ray Timing Explorer (XTE), will continue their traditional role of timing
the brighter X-ray sources well into the 1990s.

Source location, by non-imaging and imaging means, has certainly been
the preoccupation of X-ray astronomy over most of its short history. As we
have seen, grazing-incidence telescopes have been very intensively and
successfully developed over the past 15 years, but these devices do have
their scientific limitations (small fields-of-view and poor high-energy response) as well as limitations in cost. Further extension of the upper energy
cutoff of Wolter Type 1 mirrors beyond the ~ 10 keV likely to be achieved
for AXAF and XMM would involve prohibitively small grazing angles and
hence very long focal lengths in large-diameter telescopes, even if the use of
iridium in place of gold as the reflective coating improved efficiency out to
40 keV (Elvis *et al.*, 1988). Coating of short focal length mirrors with
multilayer diffractors (Catura *et al.*, 1983) might improve throughput in

Fig. 1.14. X-ray astronomy in the real world. (a) Number of XMM mirror modules as a function of calendar time. The expected launch date for XMM is 1998. (b) Scheduled AXAF launch date versus time. The history of the '1.2 metre X-ray telescope' which eventually became AXAF actually begins in the 1960s (Tucker and Giacconi, 1985).

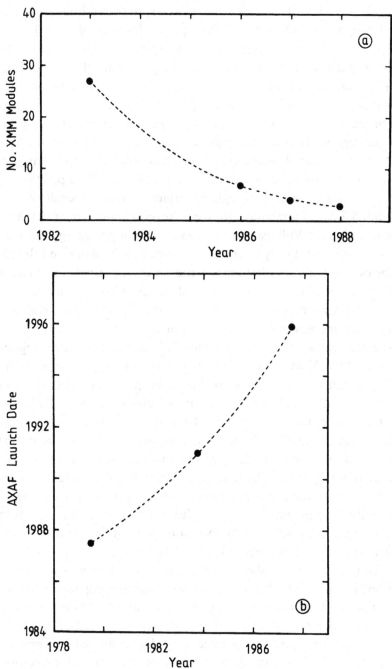

narrow bands at energies of up to 25 keV. Multilayer coatings are produced by the vacuum deposition of thin alternating layers of high (e.g. W, Au) and low (e.g. C) atomic number materials. The resulting precise periodic variation in refractive index causes the multilayer to act like a Bragg diffractor (eq. 1.3). The principal use envisaged for multilayers is in normal-incidence, narrow-band soft X-ray optics (Henry *et al.*, 1981).

The main hope for hard X-ray imaging, however, lies with coded aperture masks. In conventional imaging systems there is a one-to-one correspondence between point intensities in the object plane and in the image (detector) plane. In coded aperture imaging, incoming wavefronts are modulated (coded) by a plate containing a pattern of transparent and opaque regions; information from all points on the object then appears at each point in the detector plane. A conventional 'image' may only be reconstructed from the output of the detector (usually a position-sensitive proportional counter, or, at higher energies, a position-sensitive scintillator (section 5.2.2.2)) by recourse to one of several computer based deconvolution techniques (Willingale *et al.*, 1984). The upper energy limit of coded mask imaging is set only by the transparency of the mask and the efficiency of the detector. The earliest mask patterns were Fresnel zone plates (Mertz and Young, 1961) and random pinhole arrays (Ables, 1968; Dicke, 1968); many sophisticated patterns have subsequently been evaluated for use in astronomy and nuclear medicine (Skinner, 1984).

The first extra-solar shadowgrams (of the galactic centre region) were obtained by a University of Birmingham sounding rocket experiment in 1976 (Proctor *et al.*, 1978, 1979). The latter paper discusses the sensitivity advantages of coded mask telescopes, relative to single pinholes and scanning collimators, which arise from the simultaneous observation of many image elements. A larger, more sophisticated instrument, operating in the 3–30 keV band, was flown by the Birmingham group on the Spacelab 2 mission in July, 1985. The first images from this experiment – again of the galactic centre – have recently been published (Skinner *et al.*, 1987).

It is interesting, finally, to consider the prospects for one method of source location presently available to optical and radio astronomers but denied to X-ray astronomers – interferometry. The conceptual difficulties – interference is a wave phenomenon and X-ray astronomers are used to working in the particle (photon) picture – are more apparent than real. The principles of a photon-counting intensity interferometer, valid at any wavelength, are described by Hanbury Brown (1974). The angular distribution of intensity across a source may be measured, in principle, by observing the excess coincidence count rate between two detectors as the spacing between them is varied. The maximum separation necessary – the baseline of the interferometer – is about one metre for an angular resolution of 10^{-3} arcseconds at an X-ray wavelength of $50\,\text{Å}$ – about the separation between mirror modules in the planned XMM payload. The signal-to-noise for an intensity interferometer, however, is linearly dependent on the mean count rate in the two detectors and on the coherence time (reciprocal of the bandwidth) of the incident beam. Even with the most optimistic assumptions regarding observation times and detector time resolution, there are simply not enough source photons, in the X-ray band, per unit frequency, to ever approach an interferometric signal-to-noise ratio of unity.

As real X-ray source location techniques have matured in the past ten years, so the thrust of instrumental development has moved on, to spectroscopy. Very intensive efforts are currently being directed towards improved grating and crystal spectrometer designs and towards the development of non-dispersive, energy-resolving detectors such as CCDs and calorimeters.

If high-resolution spectroscopy is the principal concern of the late 1980s (currently there are a few tens of sources for which high-resolution spectra exist), there remains for the future the exploitation of X-ray polarimetry. Only a very small number of sensitive measurements of linear polarisation have so far been made (Novick *et al.*, 1977; Hughes *et al.*, 1984). The prospects for a radical advance in this area, however, seem rather dim. Even if the dispersive Bragg crystal and scattering polarimeters described in section 1.3 were to be combined with the large collecting area of (say) XMM, measurements of 1% linear polarisation could only be made for about the ten brightest sources (Novick *et al.*, 1985). A high-M, high-efficiency polarimeter is currently the most notable gap in the instrumental armoury that X-ray astronomers can bring to bear. Only in the recently discovered polarisation sensitivity of microchannel plates (section 3.3.8) is there any hint of how such a device might be constructed.

1.6   Detectors

No matter which observational technique is used to interrogate the X-ray sky, the outcome of a measurement is invariably determined by the properties of that electronic detector with which the collimated, focused or diffracted photons eventually interact. The remainder of this monograph describes, on a one-per-chapter basis, the classes of detector now used, or contemplated for use, in soft X-ray astronomy.

The ideal detector for satellite-borne X-ray astronomy would combine high spatial resolution with a large useful area, excellent temporal resolution with the ability to handle large count rates, good energy resolution with unit quantum efficiency over a large bandwidth. Its output would be stable on timescales of years and its internal background of spurious signals would be negligibly low. It would be immune to damage by the in-orbit radiation environment and would require no consumables. It would be simple, rugged and cheap to construct, light in weight and have a minimal power consumption. It would have no moving parts and a low output data rate.

Needless to say, such a detector – colloquially known as the 'all-singing, all-dancing' detector – does not exist. Each of the detector types we shall describe in detail in the following chapters rather possesses a small number of desirable attributes to the exclusion – to a greater or lesser degree – of all others.

Some sing. Others dance. They find completely common ground only in the amount of ingenuity that has gone into their development and in the remarkable rapidity with which they continue to evolve.

2

Proportional counters

2.1 Introduction

Gas proportional counters have been the 'workhorses' of X-ray astronomy throughout the subject's entire history. The roots of proportional counter development, however, go back much further, to the pioneering counters of Rutherford and Geiger (1908), to the first quantitative gas ionisation studies of J. J. Thomson (1899) and beyond.

The physics of gas-filled particle and X-ray detectors was very intensively researched during the four decades up to 1950. The classic texts of Curran and Craggs (1949), Rossi and Staub (1949) and Wilkinson (1950) describe a highly developed field at the zenith of its importance: before first NaI scintillators (in the 1950s) and then semiconductor detectors (in the early 1960s) replaced gas detectors in many areas of nuclear physics research.

Outside X-ray astronomy, proportional counter fortunes began to revive in the late 1960s when position-sensitive variants of the single-wire proportional counter (SWPC) were introduced as focal plane detectors for magnetic spectrographs (Ford, 1979). Multi-wire detectors (first developed, but not fully exploited, at Los Alamos as part of the Manhattan Project – Rossi and Staub, 1949) then rapidly evolved to provide an imaging capability in two dimensions over large areas. Here, the impetus was provided by the particle physicists (Charpak et al., 1968) who continue to dominate the field of gaseous detector development.

This chapter does not attempt to give a complete account of gaseous electronics, nor does it describe in detail related detector developments in fields such as particle physics (Fabjan and Fischer, 1980; Bartl and Neuhofer, 1983; Bartl et al., 1986). After a description of operational principles (section 2.2), we shall instead concentrate on the main current areas of counter development within X-ray astronomy:

(i) large-area, low-background collimated detectors (section 2.3);

(ii) imaging detectors and readout methods (section 2.4);

(iii) counters with enhanced energy resolution (section 2.5).

Examples of flight instruments, electronics and data analysis are described in each of these three final sections, together with direct applications of astronomical proportional counter research to other fields.

2.2   Physical principles of proportional counter operation
2.2.1   *X-ray interactions*

All X-ray proportional counters (fig. 2.1) consist of a windowed gas cell, subdivided into a number of low- and high-electric-field regions by some arrangement of electrodes. The signals induced on these electrodes by the motions of electrons and ions in the counting gas mixture contain information on the energies, arrival times and interaction positions of the photons transmitted by the window.

At the energies of interest here ($E \lesssim 50\,\mathrm{keV}$), X-rays mainly interact with gas molecules via the photoelectric effect, with the immediate release of a primary photoelectron, followed by a cascade of Auger electrons and/or fluorescent photons (fig. 2.2). In a multicomponent gas, for X-rays of several keV energy, the number of possible ion relaxation pathways is large. Nevertheless, the product of most primary ionisation events is a cloud of low-energy secondary electrons localised to the X-ray absorption position within the (low-field) region immediately below the window (the 'drift space'). This is because typical photo- and Auger electron ranges – the 'track lengths' over which secondary ionisation occurs – are rather short (0.1–1 mm) in common counting gases at atmospheric pressure (Smith *et al.*, 1984). While fluorescent photons may be reabsorbed at some distance from the original X-ray interaction, the relevant K- and L-shell fluorescence yields are small (fig. 2.2). Only when fluorescent photons escape completely from the gas volume is any significant fraction of the X-ray energy unavailable for the creation of secondary ion–electron pairs. Such events appear in the detector's output pulse height distribution as an 'escape peak' at a well-defined energy (fig. 2.2) below the main X-ray signal.

In principle, the relative distribution of initial X-ray interaction positions and L-shell fluorescence reabsorption positions in a xenon-filled counter contains information on the linear polarisation of the incident beam. This is because the emission of the fluorescent photons has a polarisation dependent angular anisotropy (Weisskopf *et al.*, 1985). The effect in any real counter is, however, too small to be useful. Polarimeters which either directly (Riegler *et al.*, 1970; Novick, 1974) or indirectly (Sanford *et al.*, 1970) sense the direction of photoelectron emission in a gas (which is preferentially aligned with the electric vector of the incident X-rays) have a limited sensitivity above 10 keV.

Ignoring fluorescence, the number of electrons in the initial, localised, charge cloud may be written:

$$N = E/w \qquad (2.1)$$

where *w*, the average energy to create a secondary ion pair, depends somewhat on the gas composition ($w = 26.2$ eV for argon, 21.5 eV for xenon; Sipila, 1976).

Fluctuations in N contribute to the energy resolution of the counter (sections 2.3, 2.5). Since the creation of secondary ion pairs cannot be considered as a series of mutually independent events, these fluctuations are smaller than would be expected from (random) Poissonian statistics. The variance of N, σ_N^2, is in fact given by

$$\sigma_N^2 = FN \qquad (2.2)$$

where F is the Fano factor of the gas mixture in question (Fano, 1947; Alkhazov *et al.*, 1967). F has values ranging from 0.17 (argon, xenon) to 0.32 (carbon dioxide) in common proportional counter filling gases (Sipila, 1976). Equations (2.1) and (2.2) describe equally well the statistics of X-ray interactions in solid media (see Chapter 4).

Once created, the electron cloud begins to drift towards the (nearest) anode under the influence of the applied electric field. In the case of the cylindrical SWPC, the transition between low- and high-electric-field regions is a gradual one. Electric field strength between the concentric anode (radius r_A) and cathode (radius r_C) varies with radial coordinate r (fig. 2.1a) as

$$E = (V_0/r)/[\ln(r_C/r_A)] \qquad (2.3)$$

where V_0 is the anode–cathode potential difference. In most imaging counters (figs. 2.2c–f), however, the uniform low-field drift region, where X-rays are absorbed, and the high-field 'avalanche' region near the anode(s) are geometrically distinct. If the depth of the parallel-field drift region is d, then, on exit, the (pointlike) initial charge cloud will have spread laterally (i.e. in the x or y direction) to an extent:

$$\sigma_{xy} = (2Dd/W)^{\frac{1}{2}} \qquad (2.4)$$

due to the effects of electron diffusion. Here, σ_{xy} is the standard deviation of

Fig. 2.1. X-ray proportional counter geometries. Symbols: C, cathode; A, anode; XW, X-ray window; WS, window support; ACO, anticoincidence grid; D, drift field-defining electrode; G, electron-transparent grid; UVW, ultraviolet window; COL, collimator; PSR, position-sensitive readout; d, depth of drift (X-ray absorbing) region. (a) Cross-section of cylindrical single-wire proportional counter (SWPC) (*left*) and its internal electric field (equation 2.3). Solid lines = electric field; broken lines = equipotentials. (b) Cross-section of multi-anode collimated proportional counter (section 2.3), showing cellular structure and three-sided coincidence arrangement. (c) (*left*) Imaging multi-wire proportional counter (section 2.4). A number of parallel anode wires is sandwiched between orthogonal cathode wire planes. (*right*) Electric field (solid lines) and equipotentials (broken lines) in the vicinity of adjacent anodes. Close to each anode the field of configuration resembles that of the SWPC, while several anode wire radii distant the equipotentials are parallel to one another (Tomitani, 1972; Harris and Mathieson, 1978). Borkowski and Kopp (1975) describe the 'evolution' of such MWPC structures from the single-wire counter of fig. 2.1a. Essentially, the multiple anode wires, each acting as an independent counter, are introduced between the sectioned (along the diameter aa' in fig. 2.1a) and flattened halves of the original SWPC. (d) Parallel-plate avalanche chamber (section 2.5). d' is the depth of the high-field avalanche region. (e) Imaging gas scintillation proportional counter (GSPC) (section 2.5). (f) Multi-step avalanche chamber (section 2.5).

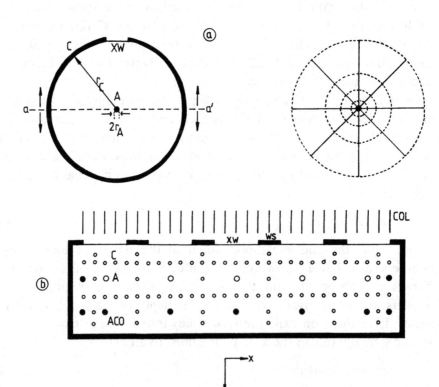

Fig. 2.1 (Continued)

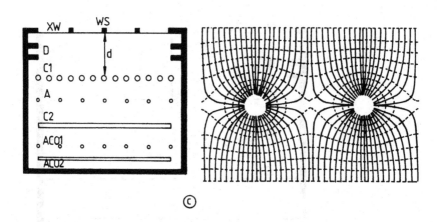

ⓒ

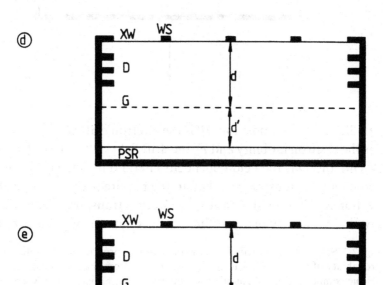

Fig. 2.1 (Continued)

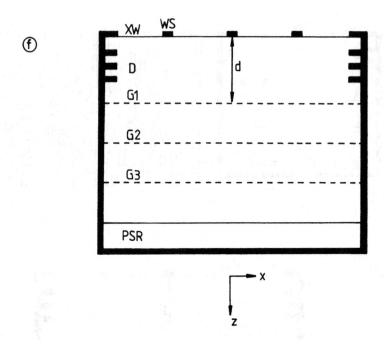

a Gaussian distribution of position, W is the electron drift velocity $(=\mu_d E_d$, where μ_d is the electron mobility and E_d the drift field strength, typically less than $1\,\mathrm{kV/cm}$) and D is the field-dependent lateral diffusion coefficient. In imaging counters, σ_{xy} must be matched to the requirements of the readout system (section 2.2.5). Calculations of the electron transport coefficients W and D have been made by many authors from solutions of the Boltzmann

Fig. 2.2 Soft X-ray interaction with a conventional proportional counter. The probability of the X-ray releasing a photoelectron from the jth shell of the ith atomic component of the counting gas mixture is p_{ij}, providing the X-ray energy E exceeds the binding energy of the shell, E_{ij}. a_i is the atomic fraction of the ith atomic type and $\sigma_{ij}(E)$ are the photoelectric cross-sections of the various shells. The excited ion may relax by either Auger electron or fluorescent photon emission. ω_{ij} is the fluorescent yield of the jth shell of the ith atomic type (Bambynek *et al.*, 1972). K-shell $(j=1)$ fluorescent yield is found empirically to vary as $Z^4/(Z^4+30^4)$, where Z is the atomic number (Zombeck, 1982). Higher shell interactions may follow the initial Auger or fluorescent emission. The fluorescent photon may escape from the counter with a probability e_{ij}, which depends on the geometry (in particular, the surface area to volume ratio) of the detector. Example: $6\,\mathrm{keV}$ X-ray absorption in the mixture Xe–10%CH$_4$. 99.83% of interactions are with the Xe L shells (mean energy $5.0\,\mathrm{keV}$). These lead to the production of a $1.0\,\mathrm{keV}$ photoelectron and a $4.2\,\mathrm{keV}$ L-fluorescence photon ($\sim 14\%$ of the time) and a $\sim 3.4\,\mathrm{keV}$ Auger electron (86% probability) since the mean M-shell binding energy in Xe is $0.8\,\mathrm{keV}$.

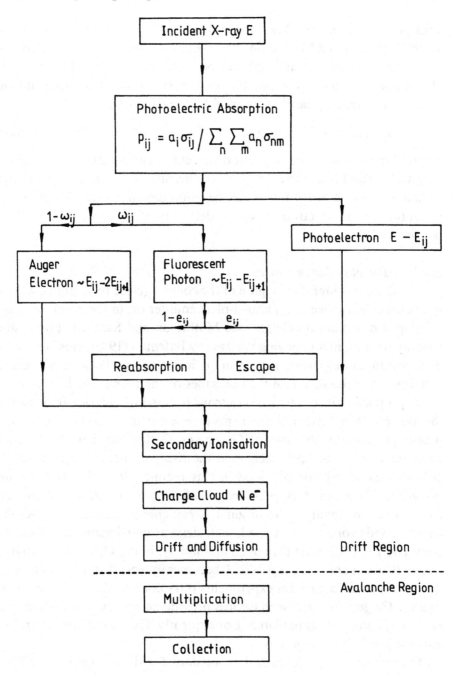

transport equation (den Boggende and Schrijver, 1984) or by Monte Carlo methods (Fraser and Mathieson, 1986). Electron diffusion is predicted, and measured (Armitage *et al.*, 1988), to be highly anisotropic. The spread of the charge cloud along the direction of drift (the *z* direction), on exit from the drift region, may be characterised by the equation.

$$\sigma_z = (2D_L d/W)^{\frac{1}{2}} \tag{2.5}$$

where, for most gas mixtures and drift field strengths, D_L/D, the ratio of longitudinal and transverse diffusion coefficients, is less than unity. Longitudinal dispersion of the initial charge cloud determines the X-ray energy resolution in the 'electron counting' detectors of Siegmund *et al.* (1982) (section 2.5.3).

2.2.2   *Multiplication processes*

Once in a high-field region of the counter (close to an anode wire or in the high-field space of a parallel-plate counter or, in the point detectors developed originally by Geiger (see Mathieson and Sanford, 1963b) and recently revived, in an imaging variant, by Bateman (1985), close to the tip of a pin anode) electrons may gain sufficient energy between successive collisions to excite or ionise the molecules of the filling gas. In a conventional proportional counter, the electrons released by collisional ionisation themselves create further electron–positive ion pairs. Electron avalanches propagate towards the anode. Gas gains *G* (the number of electrons collected at the anode per electron in the X-ray induced charge cloud) of 10^3–10^5 may readily be obtained in this manner. Provided *G* does not become too large, the final pulse magnitude, $P = NG$ electron–positive ion pairs, is proportional to the original X-ray energy via eq. (2.1). As the applied anode voltage is increased, first, strict proportionality is lost due to space charge effects near the anode and finally, in the Geiger or discharge régime (Curran and Craggs, 1949), *P* becomes independent of *E*. Although the earliest sounding rocket experiments (Giacconi *et al.*, 1962) used thin-window Geiger counters, such devices provide no energy information and have poor count rate capabilities. Consequently, Geiger counters are never now used in X-ray astronomy.

Many authors have developed expressions for the gas gain in SWPCs. The transformations necessary to render such formulae applicable to MWPCs are given by Tomitani (1972). Gas gain in a parallel-plate avalanche chamber (fig. 2.1d) is ideally $G = \exp(\alpha d')$, where d' is the depth of the avalanche region and α is the first Townsend ionisation coefficient (the number of ionisations per unit distance: Alkhazov, 1970).

The SWPC formulae reviewed by Charles (1972) and by Shalev and Hopstone (1978), all deriving from the expression due to Townsend

$$\ln(G) = -\int_{r_0}^{r_A} \alpha \, dr$$

where r_0 is the radial position of the avalanche onset and r_A is the anode wire radius (fig. 2.1a), have the form:

$$G = G[V_0/\ln(r_C/r_A), pr_A, C_i]   \qquad (2.6)$$

Here, p is the gas pressure and C_i, importantly, are constants which depend on the composition of the filling gas. It is also possible to estimate G from the microscopic collision cross-sections (elastic, vibrational, rotational, ionisation, excitation and attachment) of the gases concerned, using Monte Carlo methods (Matoba *et al.*, 1985).

A noble gas is almost always the primary constituent of the counting mixture. In X-ray astronomy, argon (for energies below $\sim 5\,\text{keV}$) and xenon (which offers good stopping power for harder X-rays: section 2.2.3) are the gases of choice. Post second world war supplies of krypton are contaminated by the radioactive ^{85}Kr isotope, a legacy of atmospheric nuclear tests. Decay of this isotope would seriously increase the internal background (B_i in eq. 1.1) of any practical detector.

The heavier noble gases, by virtue of their low w values (eq. 2.1), zero electron affinities and lack of vibrational and rotational energy loss mechanisms, would appear to be ideal X-ray detection media. Pure noble-gas-filled proportional counters, however, are extremely susceptible to trace contaminants and are unstable at high gains. The latter shortcoming mainly arises from the gas's transparency to its own UV emission. The return of a gas atom, excited during the avalanche, to its ground state may be accompanied by the emission of a UV photon of sufficient energy (~ 6–$8\,\text{eV}$ in the case of Xe) to cause photoemission from any metal surface within the counter cell. The resulting 'after-pulses' are overcome in conventional proportional counters by the addition of a polyatomic molecule (commonly carbon dioxide in X-ray astronomy) which 'quenches' the UV emission by (i) direct absorption and (ii) collisional de-excitation of the excited noble gas atom before UV emission can occur.

The light emission from a pure noble gas (usually Xe) or from certain gas mixtures (Siegmund *et al.*, 1982) can, however, be used to considerable benefit. In the gas scintillation proportional counter (GSPC), first described by Policarpo *et al.* (1972) and actively exploited by a number of X-ray astronomy groups, the primary charge cloud drifts into a region where the electric field strength has been carefully adjusted (the 'scintillation region': fig. 2.1e and sections 2.2.4 and 2.5). On average, electrons gain enough energy between collisions to excite the gas atoms to higher energy states (the first excitation potential for Xe is 8.28 eV) but not enough to ionise them (the ionisation potential for Xe is 12.1 eV). No avalanche occurs. Instead, a UV light flash, whose intensity is proportional to the initial X-ray energy, is detected by some sort of photomultiplier arrangement.

Statistical fluctuations in G are an important contributor to the energy resolution of a conventional proportional counter. The relative variance of G, $(\sigma_G/G)^2$, usually denoted by f, has values between 0.5 and 0.67 in wire chambers at low gains (Campbell and Ledingham, 1965; Sipila, 1976). In accounting for the energy resolution of GSPCs, f is replaced by a (much smaller) term related to the photon-counting properties of the UV photomultiplier (section 2.2.4). Typical GSPC energy resolutions are therefore rather better than those of conventional wire or parallel-field counters.

In all counters where avalanches propagate, the pulses observed at the anode(s) and cathode(s) are predominantly due, not to the final collection of the electrons at the active anode, but to the motion of the positive ion cloud. This follows from energy conservation considerations; because electrons, mainly generated close to the anode, traverse a relatively small potential difference before collection, the changes they can induce in the anode potential are small. Most ions, however, traverse the full anode–cathode potential difference and so gain large amounts of energy from the electrostatic field. The induced charge distributions in MWPCs, particularly, have been described by many authors (e.g. Mathieson and Harris, 1978). The negative-going anode pulse (composed of a fast (~ 1 ns) electron signal and a larger, slower ion-induced pulse lasting several microseconds) is usually used to determine the X-ray energy, while the induced charges on the cathode wire planes may be used, in a MWPC, to determine the x, y coordinates of the X-ray interaction. The final attenuation of proportional counter pulses by pulse shaping filters in the electronic signal chain is discussed by Mathieson and Charles (1969).

The avalanche, instead of surrounding the anode wire, may, depending on gas gain, be angularly localised around it (Fischer *et al.*, 1978; Harris and Mathieson, 1978; Matoba *et al.*, 1985). Some of the very many MWPC position readout methods described in section 2.4 can utilise this effect to advantage. Event positions can be interpolated, in the direction perpendicular to the anode wires, to better than the anode wire spacing, producing images which are free of anode-imposed structure.

2.2.3 Quantum detection efficiency

Irrespective of internal geometry, the quantum detection efficiency (Q in eq. 1.1) of any gas proportional counter may be written in the form

$$Q = T_w \exp(-t_w \mu_w)\{1 - \exp(-d\mu_g)\} \qquad (2.7)$$

T_w is the geometric transmission of the window support mesh which, in orbit, supports the window against the differential pressure developed across it. Typically, $T_w = 0.7$–0.8. The second term and third terms in eq. (2.7) follow from the familiar exponential attenuation law for X-rays (Zombeck, 1982). The second term is the transmission of the window, thickness t_w. The third term is the fraction of the incident beam transmitted by the window and its support structure which is subsequently absorbed in the active volume of the detector (equivalent to the drift region in most geometries: X-rays absorbed directly in the avalanche region (figs. 2.1c–f) give rise to pulses of somewhat reduced gain (Lapington et al., 1985)). $\mu_w(E)$ and $\mu_g(E)$ are, respectively, the linear absorption coefficients of the window material and of the gas. Of course,

$$\mu_w = \mu'_w \rho_w; \ \mu_g = \mu'_g \rho_g \qquad (2.8)$$

where μ'_w (μ'_g) is the mass absorption coefficient of the window (gas) and ρ_w (ρ_g) is its density.

Equations (2.7) and (2.8) tell us that in order to maximise the counter efficiency we require a window of low X-ray stopping power and a highly absorbent gas. Since photoelectric absorption, distant from any absorption edge, varies as $Z^4/E^{8/3}$ (Zombeck, 1982), this implies a window material of low atomic number Z – a plastic or light metal – and a high-pressure gas mixture of high effective atomic number. The window must be of the minimum thickness compatible with its mechanical role as a pressure seal. Plastic windows, additionally, must also be coated with a thin metallic layer (C or Al) in order to prevent optical photons entering the counter and to provide electrical conductivity.

Fig. 2.3 compares the mean X-ray absorption depths in argon and xenon at STP. We see that two atmosphere centimetres of argon provides adequate absorption ($[1-\exp(-\mu_g d)] \gtrsim 0.6$) for X-ray energies less than about 5 keV; thereafter xenon, despite its order-of-magnitude greater cost per unit volume, is the essential basis for any counting gas mixture.

X-ray proportional counters can be divided, according to their window material, into two distinct categories (see fig. 2.4):

(i) thin (plastic) windowed flow counters with a useful response to X-ray energies below 1 keV;

(ii) thick (metal) windowed sealed counters limited to energies above 2 keV.

In the latter type of counter, the window, commonly ∼50 μm of beryllium, acts as a completely gas-tight seal; the counter volume is filled on a once-and-for-all basis. In the former type, an auxiliary reservoir, gas control and flushing system, which may add considerably to experiment weight and complexity, replaces gas lost through the permeable submicron plastic membrane. Leak rates of xenon through plastic foils are described by Smith *et al.* (1987). In multicomponent counting gases, the leakage of

Fig. 2.3. Mean absorption depth $1/\mu_g$ as a function of X-ray energy in argon and xenon at STP. Gas densities 1.78×10^{-3} and 5.9×10^{-3} g/cm³, respectively. The vertical lines indicate the K absorption edge of argon (3.21 keV) and the L edges of Xe (4.78, 5.0 and 5.4 keV). The K edge of xenon lies at 34.65 keV.

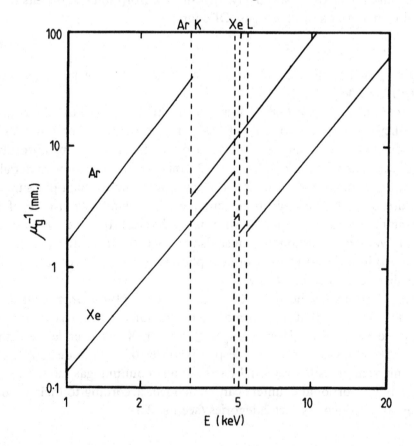

different components at different rates (argon faster than methane, for example) may cause the gas gain (and hence the energy-scale calibration of the counter) to change with time (eq. 2.6). Active gain control systems which automatically raise the anode high voltage in response to the signal from an on-board radioactive calibration source may be used, as with the

Fig. 2.4. Standard X-ray windows. Transmission $100[\exp(-\mu_w t_w)]$ as a function of X-ray energy. (1) 1 μm polypropylene (CH_2). (2) 1 μm Lexan (the polycarbonate $C_6H_{14}O_3$). (3) 1 μm parylene N (C_8H_8). Note the narrow passband associated with the C K absorption edge in all these plastics, density ~ 1.1 g/cm^3. (4) 50 μm Be ($\rho = 1.85$ g/cm^3). (5) 50 μm Al ($\rho = 2.7$ g/cm^3). Note the influence of the Al, O and C K edges at 1.56, 0.53 and 0.28 keV, respectively. Additional mass absorption data is tabulated by Zombeck (1982) for teflon (CF_2), mylar $(C_{10}H_8O_4)$ and carbon. Lexan is a trademark of General Electric Plastics. Parylene N is a trademark of Union Carbide Corporation.

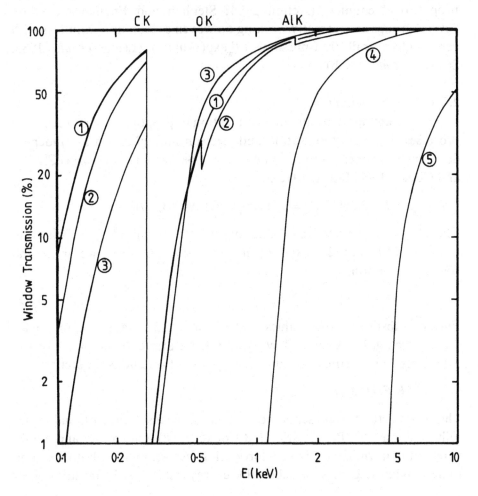

EXOSAT parallel-plate proportional counters (section 2.4.3: Mason *et al.*, 1984), to compensate for the effects of differential diffusion.

The problems of producing pinhole-free large-area beryllium foils are described in some detail in section 2.3, in the context of the Ginga Large Area Counter. Submicron polypropylene windows are generally produced by biaxially stretching much thicker (~ 25 μm) commercial stock, while applying heat and monitoring interference fringes to gauge thickness. At least at Leicester, this is a manual process with a low success rate and high frustration factor. Lexan pellicles, which also have useful properties as passband filters for XUV astronomy (section 3.3.6), are produced by casting on water (Huizenga *et al.*, 1981). Robust proportional counter windows 0.15 μm thick and up to 65 mm in diameter have been produced by this method (Bleeker *et al.*, 1980). Pinhole-free Lexan–polypropylene composite windows have recently been developed for the ROSAT imaging proportional counters (section 2.4.4: Stephan and Englhauser, 1986). Several groups are currently working on woven carbon fibre windows with very high overall transmissions ($T_w \exp(-\mu_w t_w)$) (Leake *et al.*, 1985; Schwarz *et al.*, 1985b).

2.2.4   *Energy resolution*

Assuming that, in any conventional proportional counter, the processes of primary ionisation and electron multiplication are independent, the fwhm energy resolution $\Delta E/E$ may be written in the form (Charles and Cooke, 1968; Sipila, 1976):

$$\Delta E/E = 2.36(\sigma_p/P) = 2.36\{(\sigma_N/N)^2 + N^{-1}(\sigma_G/G)^2\}^{\frac{1}{2}}$$

where $(\sigma_r/P)^2$ is the relative variance of the pulse magnitude P. Substituting from eqs. (2.1) and (2.2), and again denoting $(\sigma_G/G)^2$ by f, we obtain the standard equation:

$$\Delta E/E = 2.36\{w(F+f)/E\}^{\frac{1}{2}} \tag{2.9}$$

Further substituting the values $f = 0.67$, $w = 26.2$ eV appropriate to an argon-based gas mixture at low gas gains, we arrive at the origin of the 'workhorse' proportional counter energy resolution quoted in section 1.3:

$$\Delta E/E \sim 0.35/E^{\frac{1}{2}}$$

This figure actually represents a lower limit to the resolution attainable in a fully optimised SWPC (Charles and Cooke, 1968). Gas purity and anode wire non-uniformity are two compromising factors: loss of photoelectrons to the window degrades the achievable energy resolution (by perturbing N)

at low X-ray energies. In any large-area or imaging proportional counter, mechanical non-uniformity (variations in anode diameter and anode–anode spacing in MWPCs; variations in anode–grid spacing in parallel-plate avalanche chambers (Stumpel *et al.*, 1973)) is a serious source of energy blur. One must therefore distinguish between 'spot' and 'full-field' energy resolutions measured with a finely collimated X-ray beam and with uniform counter illumination, respectively. Full-field energy resolution is always the poorer.

Three physical parameters determine the energy resolution of a conventional (avalanche) counter (eq. 2.9); the mean ion pair energy w, the Fano factor F, and f, the relative variance of the gas gain. Three routes towards improved energy resolution have been actively pursued by X-ray astronomers. These are (section 2.5):

(i) Use of Penning gas mixtures (Penning, 1934) with low values of F and w (Sipila, 1976; Schwarz and Mason, 1984; Sephton *et al.*, 1984).

(ii) Use of advantageous multiplication geometries. In parallel-plate avalanche chambers (fig. 2.1d), f has the ideal weak-field value of unity. For certain optimum field strengths, f can be made to approach zero in a parallel-plate geometry (Alkhazov, 1970). This has led to the recent development of multistep avalanche chambers, in which the necessary gas gain for imaging is obtained in two separate avalanche regions (fig. 2.1f: Ramsey and Weisskopf, 1986). The Penning Gas Imager (PGI) proposed by Schwarz and Mason at MSSL in fact combines a multistep chamber geometry with a Penning gas filling, argon–acetylene.

(iii) Elimination of the avalanche variance either by detecting the integral UV light yield from the primary electrons drifting in a scintillation region, as in a GSPC, or by 'counting' the time-resolved light flashes from individual primary electron avalanches (Siegmund *et al.*, 1982). For a gas scintillation proportional counter, the counterpart to eq. (2.9) is (Simons *et al.*, 1985a):

$$\Delta E/E = 2.36\{wF/E + 1/N_p\}^{\frac{1}{2}} \qquad (2.10)$$

where N_p is the average number of photoelectrons, per incident X-ray, produced in the UV photomultiplier. As N_p tends to infinity in a xenon GSPC one obtains a limiting resolution

$$\Delta E/E = 0.14/E^{\frac{1}{2}}$$

more than a factor of two better than that of a conventional counter.

2.2.5 *Spatial resolution*

For any imaging proportional counter at the focus of a grazing-incidence X-ray telescope, the fwhm position resolution Δx – the uncertainty in determining the x coordinate of an X-ray interaction – may be written as the quadratic sum of at least four independent contributions:

$$\Delta x = 2.36\{(\Delta x_r)^2 + (\Delta x_d)^2 + (\Delta x_t)^2 + (\Delta x_n)^2\}^{\frac{1}{2}} \qquad (2.11)$$

The first two terms represent the fundamental gas physics limits to resolution. Δx_r is the contribution of the photoelectron (or Auger electron) range, which, according to Gilvin *et al.* (1981a) and Smith *et al.* (1984), may be written in the form

$$\Delta x_r = aE_e^n/\rho_g \ \mu m$$

where E_e denotes the electron energy, ρ_g is the gas density and a and n are constants ($a = 30$, $n = 1.3$ for $1 < E < 5$ keV). For 6 keV X-rays interacting with xenon at STP (fig. 2.2), Δx_r is thus about 25 µm. We see that Δx_r and ρ_g are inversely related. Thus, achieving the ultimate spatial resolution in a gas counter requires high-pressure operation. The results of Smith *et al.* (1985b) and Fischer *et al.* (1986) (X-ray resolutions of 22 and 14 µm fwhm in xenon–CO_2 mixtures at 5 and 10 bar) have not yet been duplicated in any astronomical counter.

Secondly, Δx_d is the contribution of lateral diffusion in the drift space of the counter. According to Bleeker *et al.* (1980), if the effects of binning by the readout (Gilvin *et al.*, 1981a) can be ignored, Δx_d in a conventional (avalanche) counter has the form:

$$\Delta x_d = (\sigma_{xy}/N^{\frac{1}{2}})\{1 + (\sigma_G/G)^2\}^{\frac{1}{2}} = (\sigma_{xy}/N^{\frac{1}{2}})(1+f)^{\frac{1}{2}} \sim \sqrt{2}\sigma_{xy}/N^{\frac{1}{2}} \qquad (2.12)$$

Each of the N electrons in the initial charge cloud (eq. 2.1) is taken to initiate a separate avalanche. It is the mean position of these N individual avalanches which is always registered by the readout (section 2.4). Diffusion effects tend to be important at low X-ray energies, where N is small. Equations similar to (2.12) can be derived for imaging GSPCs (Simons *et al.*, 1985b: see section 2.5.2).

Thirdly, Δx_t is the result of X-ray penetration into the gas. X-rays from a grazing-incidence telescope converge to their focus along the surface of a cone (half-angle θ_t, say). Absorption in the gas (section 2.2.3) therefore takes place along a path inclined at an angle θ_t to the detector's z-axis. For a counter with its window positioned at the telescope focus, this 'parallax' term has magnitude

Fig. 2.5. One-dimensional Monte Carlo simulation of the effects of X-ray penetration in a parallel-plate avalanche chamber (Sanford *et al.*, 1979). Gas mixture = 80% argon/20% methane, 1.1 bar. Absorption depth $d = 3$ mm. Perfect X-ray telescope, cone angle $\theta_t = 8°$. (a) Distribution of absorption positions with counter window placed at the telescope focus. X-ray energies 0.25, 0.5, 1.0 and 2.0 keV. (b) As (a), except that the focus is displaced 0.75 mm into the drift (absorption) region of the counter. Now at low energies the image of a point source is an annulus; at higher energies, however, fewer counts lie in the broad wings of the distribution. (*Courtesy P. Sanford. © 1979 IEEE.*)

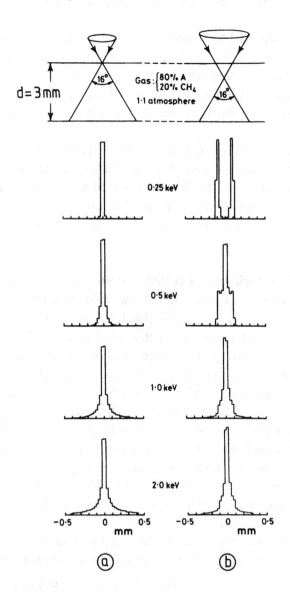

$$\Delta x_t \sim \sin\theta_t / \mu_g$$

X-ray penetration effects usually dominate the overall spatial resolution of imaging proportional counters at X-ray energies greater than about 5 keV. Telescope cone angles vary from $\sim 2°$ (XMM) to $\sim 7°$ (ROSAT). Xenon is obviously a better basis for the gas mixture than argon in this context. It is often advantageous to deliberately offset the telescope focus from the window plane into the drift region (fig. 2.5: Sanford *et al.*, 1979).

Finally, Δx_n is the result of finite signal-to-noise in the (as yet unspecified) position readout element. The physical origins of this term depend on the encoder type. Specific expressions for Δx_n are given in section 2.4, where readout technique in gas counters is reviewed in detail. A fifth term – due to partition noise – may enter eq. (2.11) depending on the type of readout used.

In all imaging X-ray detectors, spatial resolution is usually measured either from the image profile of a pinhole in an illuminated test mask, or from the transition from 'light' to 'dark' at an edge. These methods, together with a method of wire shadows, which determines resolution from the depth of shadow cast by an opaque linear obstruction, are described by Fraser *et al.* (1988a).

2.3   Large-area, low-background collimated proportional counters

Chapter 1 described in some detail the role of large-area gas counters in X-ray astronomy, and highlighted the evolution of these instruments towards larger and larger geometric areas (see also Table 2.1). For any non-imaging counter, point source detection sensitivity scales only as $A_s^{\frac{1}{2}}$, the square root of the counter area (eq. 1.1) and is severely limited (to about one-thousandth of the flux of the Crab Nebula) by source confusion in the $\sim 1°$ field-of-view of its collimator. Counter sensitivity to temporal variations, on the other hand – the ability to detect X-ray bursts, dips and periodicities and the recently discovered 'quasi-periodic oscillations' in source intensity – scales linearly with collecting area. Non-imaging proportional counters thus continue to be developed for timing (and low-resolution spectroscopy) of the brightest celestial sources.

This section describes the construction, calibration and operation of large-area proportional counters with reference to two recent experiments:

  (i) the EXOSAT Medium Energy Detector Array (MEDA) (Taylor *et al.*, 1981; Turner *et al.*, 1981); and

(ii) the Large Area Counter (LAC) on the Ginga Satellite (Makino *et al.*, 1987).

2.3.1 Construction and calibration

ESA's EXOSAT observatory is shown schematically in fig. 2.6. The MEDA, an array of eight 'double cell' proportional counter modules, is mounted, coaligned with the remainder of the X-ray instrumentation, on one side of the three-axis-stabilised spacecraft. The Large Area Counter, also consisting of eight modules, occupies a similar position on the Ginga Satellite, launched by a Mu 3SII rocket into low earth orbit on 5 February, 1987. Ginga ('Milky Way') is the third Japanese national X-ray astronomy satellite, known pre-launch as Astro C. Its predecessors were: Hakucho ('Cygnus'; launched 21 February, 1979, re-entered April, 1985) and Tenma ('Pegasus'; launched 20 February, 1983, ceased operations December, 1985).

EXOSAT's MEDA was originally proposed to ESA by the X-ray Astronomy Group at Leicester, the Max Planck Institut at Garching and the University of Tübingen in 1973. The scientific characteristics of the

Table 2.1. *Non-imaging gas counter geometric areas A_s (eq. 1.1)*

Year	Experiment	A_s (cm²)	Mode
1962	AS&E/MIT sounding rocket (Giacconi *et al.*, 1962)	20	S
1968	Leicester Skylark SL723 (fig. 1.2)	3000	S
1974–80	Ariel V Sky Survey Instrument (Villa *et al.*, 1976)	280	S
1977–9	HEAO-1 Experiment A1 (Friedman, 1979)	11 900	S
1983–6	EXOSAT Medium Energy Detector Array (Turner *et al.*, 1981)	2000	P
1987–	Ginga Large Area Counter (Makino *et al.*, 1987)	4100	P
?	X-ray Timing Explorer (Bradt, 1982)	10 000	P
?	Stanford–NRL X-ray Large Array (Gursky, 1986)[a]	1 000 000	P

[a] The 1–100 keV X-ray Large Array has been proposed as part of the US space station programme. In addition to its astronomical role, it would act as a sensitive monitor for the presence of radioactive materials in near-earth space.
Operational modes: S = scanning (sky survey); P = pointing (timing).

MEDA are summarised in Table 2.2. Many of the novel constructional features of this instrument were adopted in response to rather severe spacecraft constraints on instrument mass and envelope. Such accommodation constraints inevitably accompany the transfer of detector techniques from the laboratory, where resources are plentiful, to orbit, where mass, volume, telemetry and electrical power are always at a premium. In

Fig. 2.6. Exploded view of the EXOSAT observatory (Taylor *et al.*, 1981). EXOSAT was launched into a highly elliptical (191 708 × 347 km, 90.6 hr) deep space orbit by an American Thor Delta 3914 rocket on 26 May, 1983, re-entering the Earth's atmosphere on 6 May, 1986. The total spacecraft mass was 500 kg, of which the instrumental payload accounted for 120 kg. (*Courtesy B. Taylor.* © *1981 by D. Reidel Publishing Company. Reprinted by permission of Kluwer Academic Publishers.*)

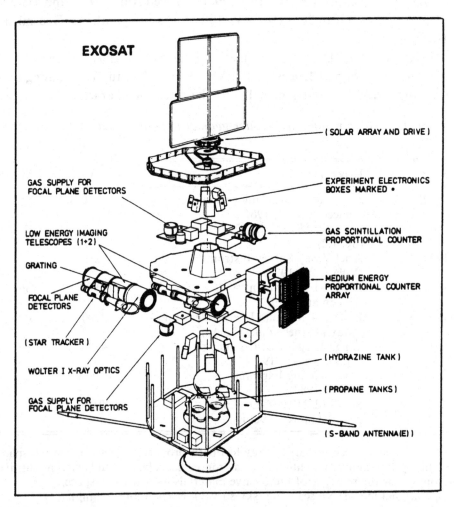

EXOSAT

(SOLAR ARRAY AND DRIVE)

GAS SUPPLY FOR FOCAL PLANE DETECTORS

EXPERIMENT ELECTRONICS BOXES MARKED •

LOW ENERGY IMAGING TELESCOPES (1+2)

GAS SCINTILLATION PROPORTIONAL COUNTER

GRATING

MEDIUM ENERGY PROPORTIONAL COUNTER ARRAY

FOCAL PLANE DETECTORS

(STAR TRACKER)

(HYDRAZINE TANK)

WOLTER I X-RAY OPTICS

(PROPANE TANKS)

GAS SUPPLY FOR FOCAL PLANE DETECTORS

(S-BAND ANTENNA(E))

order to reduce mass, the MEDA detector bodies, like the bodies of the Uhuru proportional counters before them (Tananbaum and Kellogg, 1970), were machined out of solid beryllium – a low-density, thermally stable but highly toxic metal. For the same reason, the MEDA collimators (and those of the EXOSAT GSPC: section 2.5.1.2) were produced from lead glass by the same drawing and fusing techniques used in microchannel plate manufacture (section 3.2). Each collimator element was $35 \times 47 \, \text{mm}^2$ in area and only 11 mm deep, with 150 μm square channels separated by 30 μm septa (see fig. 1.3). A mosaic of 24 such elements equipped each detector module. By these and other techniques individual module masses were kept down to 3.5 kg.

The eight proportional counter modules of the MEDA (labelled A–H) were grouped in four pairs (quadrants) (fig. 2.6) and mounted on a tilt mechanism. Each quadrant could then be independently orientated and the four fields-of-view offset one from another. Originally designed to provide a small angular region of uniform response for lunar occultation measurements, this mechanism could also be used to obtain simultaneously source and source plus background count rates by offsetting half the array to a blank field during long (1–$4 \times 10^4 \, \text{s}$) pointed observations. Typically, the 'source' and 'background' halves of the array were exchanged ('nodded') during each observation in order to account for small systematic differences in B_i between individual modules. The total mass of the MEDA was 48 kg.

Each of the eight modules was sealed by a thick (62 μm (three quadrants) or 37 μm (one quadrant)) beryllium X-ray window, giving a low-energy

Table 2.2. *Scientific characteristics of the EXOSAT Medium Energy Detector Array*

Total geometric area (A_s)	$\sim 2500 \, \text{cm}^2$ (all eight modules aligned)
Total effective area (QA_s)	$\sim 1500 \, \text{cm}^2$
Bandwidth (δE)	1.5–20 keV (argon cells)
	5–50 keV (xenon cells)
Full-field energy resolution ($\Delta E/E$)	$\sim 51/E^{\frac{1}{2}}\%$ fwhm (argon)[a]
	18% fwhm (xenon) (10–30 keV)
Collimator transmission ($\theta_{\frac{1}{2}}$)	45' fwhm (triangular response has a 3' flat-top)
Total internal background after rejection (B_i)	~ 3 counts/keV s (2–10 keV) argon
	~ 12 counts/keV s (5–50 keV) xenon

[a] The MEDA was in fact only 'semi-proportional', so that the $E^{-\frac{1}{2}}$ law was not fully obeyed.

cutoff at around 1.5 keV. Internally, each module had a 'double cell' structure, the 8 cm active depth being divided in two by a 1.5 mm thick internal window, also made from beryllium. The upper cell, pressurised with an argon–CO_2 mixture to 2 bar, contained two planes of anode wires with an interleaved cathode plane. The lower chamber, filled with Xe–CO_2 to the same pressure, contained three anode planes. Each MEDA module thus performed the function of two separately optimised multi-anode proportional counters, the lower Xe cell extending the overall sensitivity to about 50 keV (see fig. 2.7).

By comparison with the MEDA, Ginga's Large Area Counter (LAC) (fig. 2.8) may seem at first sight a technically unambitious instrument, distinguished only by sheer size. The eight permanently coaligned modules of the LAC, with their conventional stainless steel construction and 88 kg mass, have a total geometric area in excess of 4000 cm² (Table 2.1 and fig. 2.9), the largest of any pointed (as opposed to scanning) proportional counter so far flown on a satellite. 'Scaling-up' from the MEDA, however,

Fig. 2.7. EXOSAT MEDA count rates during a calibration observation of the Crab Nebula. For X-ray energies above 10 keV, the lower (xenon) cells of the detector give a higher count rate than the upper (argon-filled) chambers (Parmar and Smith, 1985). (*Courtesy A. Smith.*)

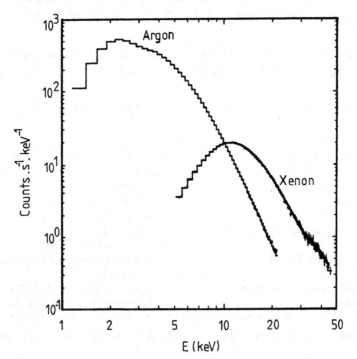

presented its own formidable technical problems, particularly with regard to the production of the LAC multi-wire arrays (MWA) and beryllium windows.

Fig. 2.10 outlines the manufacture and testing of the LAC modules. Note the emphasis on (i) Cleanliness – most assembly operations were carried out under 'clean room' conditions. (ii) Electrical-breakdown-free high voltage operation. Anode voltages of ~ 1830 V were required to produce gas gains of ~ 2000. The gas gain was kept deliberately low in order to minimise the avalanche variance f and hence maximise energy resolution (sections 2.2.2, 2.2.4). (iii) Pressure integrity. To pass its final leak-test, each detector module had to be leak tight to better than 10^{-8} Torr l/s, when pressurised with 2 bar He. (iv) Vibration ('shake') tests, to qualify the counters, and in particular the MWAs, for the launch environment.

Each LAC module had an internal gas volume 56 cm long by 18 cm wide by 5.6 cm deep, divided lengthwise into cells, as shown schematically in figs. 2.1b and 2.12, by the 52 anodes and 396 cathodes of the MWA. The anode wire diameters were 38 μm, those of the cathodes 50 μm. The total length of wire in each array was 226.7 m. Problems with any one wire could, and, during development, did, require the entire array to be restrung.

Fig. 2.8. Ginga Large Area Counter (LAC) module. Uppermost section – collimator block. Central section – gas cell. Lowest section – flight electronics box. (*Courtesy H. D. Thomas.*)

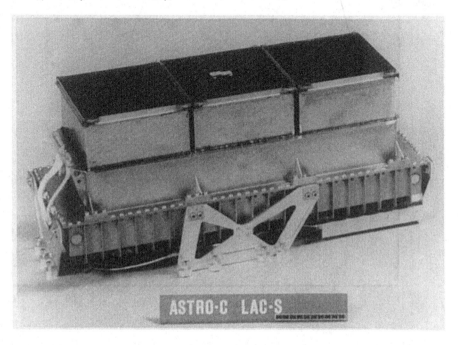

Each LAC window assembly contained a mosaic of six $15 \times 7.5\,\text{cm}^2$, $62.5\,\mu\text{m}$ thick beryllium foils. The design was originally based on 15 cm square foils, larger than those in the MEDA, but the yield of gas-tight windows then proved unacceptably (and unexpectedly) low. Ultrapure beryllium is a brittle metal; the rolling of thin sheet produces a certain

Fig. 2.9. LAC effective area as a function of X-ray energy (all eight modules). Each module is filled with $75\%\text{Ar} + 20\%\text{Xe} + 5\%\text{CO}_2$ at a pressure of ~ 2 bar. Active depth $d = 4\,\text{cm}$, Be window thickness $t_w = 62.5\,\mu\text{m}$, giving a bandpass of 1.5–35 keV. The curve shown here includes secondary effects such as transmission through the spacecraft thermal protection foil and, above the Xe K edge, the effects of K-shell fluorescence losses. (*Courtesy H. D. Thomas.*)

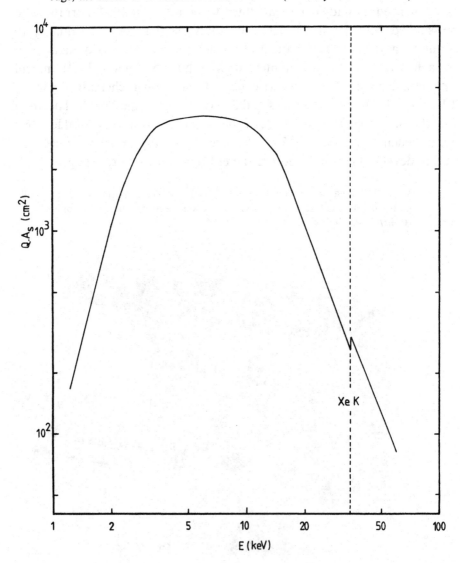

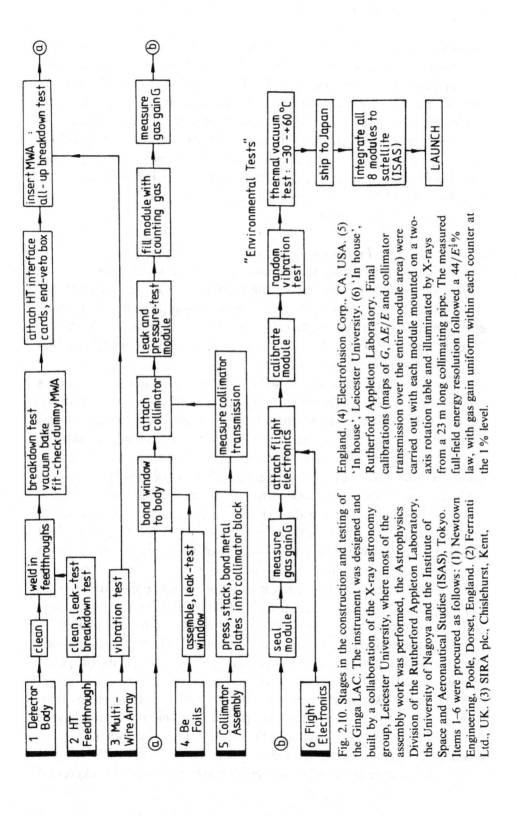

Fig. 2.10. Stages in the construction and testing of the Ginga LAC. The instrument was designed and built by a collaboration of the X-ray astronomy group, Leicester University, where most of the assembly work was performed, the Astrophysics Division of the Rutherford Appleton Laboratory, the University of Nagoya and the Institute of Space and Aeronautical Studies (ISAS), Tokyo. Items 1–6 were procured as follows: (1) Newtown Engineering, Poole, Dorset, England. (2) Ferranti Ltd., UK. (3) SIRA plc., Chislehurst, Kent, England. (4) Electrofusion Corp., CA, USA. (5) 'In house', Leicester University. (6) 'In house', Rutherford Appleton Laboratory. Final calibrations (maps of G, $\Delta E/E$ and collimator transmission over the entire module area) were carried out with each module mounted on a two-axis rotation table and illuminated by X-rays from a 23 m long collimating pipe. The measured full-field energy resolution followed a $44/E^{\frac{1}{2}}\%$ law, with gas gain uniform within each counter at the 1% level.

density of cracks per unit area. Thus, if the size of the foil required is increased, the yield of pinhole-free foils must inevitably fall, becoming uneconomic somewhere beyond the $15 \times 7.5\,cm^2$ format. The cost of beryllium procurement for the LAC windows was about £100000.

Finally, note that as well as being toxic and brittle, beryllium is also attacked by water vapour. In orbit failure of part of the Ariel V SSI (Villa *et al.*, 1976) was attributed to high launch site humidity causing a slow post-launch leak through the 85 µm thick beryllium window. Tananbaum and Kellogg (1970) describe in detail the corrosion testing of beryllium windows and counter bodies.

2.3.2   *Background rejection*

For any non-imaging detector, the necessary accompaniment to large area is a low internal background count rate B_i (eq. 1.1). Satellite X-ray astronomy is commonly conducted in two quite different radiation environments: (i) in low earth orbit, at altitudes less than 600 km (e.g. Uhuru, Ariel V, ROSAT, AXAF), or (ii) in deep space, at orbital altitudes up to 200000 km (e.g. EXOSAT, XMM).

In the latter environment, outwith the shadow region of the earth's radiation belts, particle fluxes vary (by factors of two to three) with the cycle of solar activity. The isotropic flux of $\sim$ GeV interplanetary cosmic rays (protons with a small percentage of α particles) accounts for most of the ~ 1 count/cm^{2}s raw background rate observed with gas counters at solar minimum (Mason and Culhane, 1983; Briel *et al.*, 1988).

Count rates observed in low earth orbit vary strongly with the geomagnetic latitude, and hence the orbital inclination, of the spacecraft. A detailed analysis of the raw background has been performed by Mason and Culhane (1983). Among the major sources of background events are (i) primary cosmic rays, (ii) 'albedo' electrons with energies in excess of 10 MeV, resulting from cosmic ray interactions in the upper atmosphere, (iii) high-energy photons, which usually produce, via Compton scattering, a fast recoil electron, (iv) electrons trapped in the earth's magnetic field, and (v) radioactivity induced in the detector, detector shielding and spacecraft structure by passage through the radiation belts or South Atlantic Anomaly (Dyer *et al.*, 1980).

While raw background count rates in low earth orbit are an order of magnitude less than in interplanetary space, they remain comparable with the X-ray fluxes of even the brightest cosmic sources (section 1.1). Non-imaging X-ray astronomy, therefore, would be barely possible without efficient rejection techniques to distinguish between unwanted background events and the desired X-ray signal. Background rejection techniques in all gas counters fall into three distinct categories:

(i) energy selection – rejecting all events which deposit energies outside the X-ray bandpass;

(ii) rise-time discrimination (RTD);

(iii) anti-coincidence techniques within a sub-divided gas cell. This form of anti-coincidence has now wholly superceded the practice, reviewed by Cooke *et al.* (1973), of surrounding the active gas volume with a shield of plastic scintillator.

The usefulness of energy selection depends on the observed energy spectrum of the raw background and, hence, on the nature of the particle flux and the geometry of the detector gas cell. In practice, the reduction of B_i by the hundred-fold factor which is usually sought requires the use of techniques (ii) and (iii), both of which distinguish between pointlike X-ray events and the extended tracks of minimum ionising particles.

Rise-time discrimination, first developed at Leicester by Mathieson and Sanford (1963a) and implemented soon thereafter in sounding rocket detectors (Cooke *et al.*, 1967), distinguishes between X-ray and particle events with the same initial number of ion pairs on the basis of anode pulse shape. Pulses produced in the gas by minimum ionising particles generally have slower rise-times than those produced by kiloelectronvolt X-rays. Fig. 2.11a is an early laboratory demonstration of RTD (Culhane *et al.*, 1966). Here, fast Compton electrons produced by γ-ray interactions in the metal counter walls were used to irradiate the gas volume. Cosmic ray muons are the other 'standard' calibration source in background rejection experiments.

The electronic circuitry required to implement RTD is shown schematically in fig. 2.11b. The efficiency of the rise-time technique is found to be a decreasing function of X-ray energy (Harris and Mathieson, 1971). As photoelectron ranges increase, the distinction between 'point' and 'extended' events disappears. For 5.9 keV X-rays, rejection efficiencies of 97% (background reduction factors 30:1), coupled to X-ray acceptances of 95% (5% true signal loss) can be achieved (Gorenstein and Mickiewicz, 1968; Harris and Mathieson, 1971; Culhane and Fabian, 1972).

The analogue of RTD in gas scintillation proportional counters is *burst length discrimination* (Andresen *et al.*, 1977a). Here, the duration (and profile) of the UV light flash is taken as an index of the spatial extent of the initial event; rejection efficiencies in excess of 97% for ^{60}Co γ-ray counter irradiation have been reported in GSPCs (section 2.5.1.2).

The third technique of background rejection is the use of independent 'guard' and 'veto' counters, operated in anti-coincidence within a common gas cell (figs. 2.1b and 2.12). Fig. 2.12 shows the various anode wire groups within each module of the Ginga LAC. Coincident pulses on any two groups, indicating an extended source of ionisation, cause the event to be rejected. The central gas cell (anodes S23) is, in fact, guarded on all six sides

Fig. 2.11. (a) Background rejection by rise-time discrimination (RTD) (Culhane *et al.*, 1966). Filled circles = pulse height distribution, proportional counter irradiated with 662 keV ^{137}Cs γ-rays and 5.9 keV ^{55}Fe X-rays, no discrimination applied. Open circles = identical irradiation, slow-rising pulses rejected electronically. (*Courtesy J. L. Culhane.*) (b) Block diagram of RTD electronics (Harris and Mathieson, 1971). The amplified voltage pulse from the proportional counter anode is shaped by a doubly differentiating, singly integrating filter. The zero-crossing time of the bipolar waveform at the filter output is a measure of the rise-time of the anode pulse. Signals from a time-to-amplitude converter (TAC) are then used to gate acceptance of each event, according to whether or not its zero-crossing time lies within the limits of previously measured X-ray rise-time spectra. (*Courtesy T. J. Harris.*)

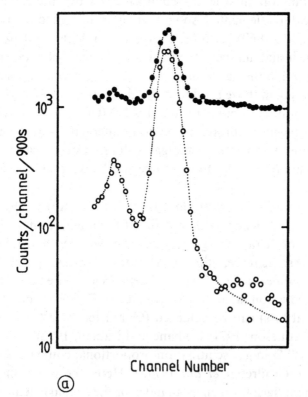

(a)

Fig. 2.11. (Continued)

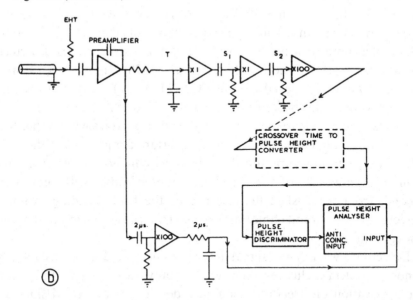

(b)

against entrant particle tracks. Anode groups V1 and V2 together form the traditional three-sided guard (fig. 2.1b) against photon-induced events originating in the counter walls, while groups R1 and L1 make up a front guard against particles obliquely entering via the X-ray window. 'End-veto' electrodes at the ends of the MWA complete the six-sided arrangement which, without the use of RTD, reduce the background to only 12 counts/s per LAC module in the 500 km Ginga orbit. The loss of genuine X-ray

Fig. 2.12. Cross-section of Ginga multi-wire array, showing interconnected anode groups L1, R1, S23, V1, V2. Open circles represent positions of individual cathode wires. Each wire in the array is 50.6 cm long.

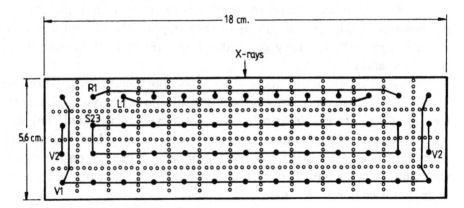

events, due to photoelectron tracks crossing anode cell boundaries, is estimated to be less than $\sim 10\%$ at all energies in the 1.5–35 keV bandpass.

The five-sided anti-coincidence geometry of the EXOSAT MEDA, which incorporated a novel end-guard cathode structure, is described by Turner *et al.* (1981) and by Bailey *et al.* (1978). In the laboratory, the MEDA counters achieved overall ^{60}Co (1173 keV) γ-ray and cosmic ray muon rejection efficiencies of 99% for 95% 2–6 keV X-ray acceptance. In deep-space orbit, where the irradiating particle energies were rather higher, the rejection rate rose to 99.6%. One important component of the residual MEDA background, uranium L-shell X-ray emission, arose from unwelcome contamination of the beryllium counter bodies with plutonium; a second, from the L-shell fluorescence of the lead shielding around the detector array. Careful choice of detector materials is essential if minimum background is to be achieved.

Detecting a weak X-ray signal in the presence of a dominant background is not a problem exclusive to X-ray astronomy. Leake *et al.* (1985) describe the application of rejection techniques developed for the MEDA to the monitoring of plutonium in the lungs of workers in the nuclear industry. The external signature of plutonium inhalation is a weak flux of uranium L series (13.6–20 keV) X-rays which may be registered using any one of a number of detector types. Rapid detection of small intakes, however, requires effective discrimination against the natural radioactivity of the human body. A prototype proportional counter monitor developed by the UK Atomic Energy Research Establishment, Harwell, and the X-ray Astronomy Group at Leicester has been shown to be more sensitive than the phoswich detectors (section 5.2.1) conventionally used for Pu-in-lung screening.

2.3.3   Lifetime

The influence of the in-orbit particle background on counter performance is both instantaneous (in creating noise) and cumulative, in that damage to the counting gas and anode wires may result from long-term irradiation.

Methane is an attractive quench gas for proportional counters because of the very high electron drift velocities measured in argon– and xenon–methane gas mixtures. Nevertheless, carbon dioxide is preferred in almost every astronomical application because of the tendency of hydrocarbons to polymerise ('crack') under irradiation, producing a deposit on the anode wire which changes its effective radius. Changes in gas gain result from these increases in r_A (eq. 2.6).

The finite lifetime of methane proportional counters has been measured by several groups (den Boggende *et al.*, 1969; Smith and Turner, 1982). The latter authors established the following limits to safe ^{60}Co exposure during the EXOSAT MEDA development programme:

Ar–CH_4: $<10^{-5}$ C (abstracted anode charge)/mm anode length
Xe–CH_4: $<5 \times 10^{-7}$ C/mm

The most dramatic manifestation of methane cracking occurred in the development of the high-field parallel-plate avalanche chamber for EXO-SAT, where high-energy background events were found to be capable of initiating highly destructive sparks (Cockshott and Mason, 1984: see section 2.4.3). These sparks promoted carbon deposition, spurious low-amplitude pulsing and, eventually, continuous detector breakdown. Even with CO_2-based gas mixtures, however, heavily ionising particles, generating charges 10^4–10^5 times greater than those of kilovolt X-rays, may cause anode–cathode sparks of sufficient violence to destroy $10\,\mu m$ diameter tungsten anode wires (Pfeffermann and Briel, 1985).

No long-term irradiation data has yet been obtained with the Penning gas Ar–C_2H_2 or with the other gases used experimentally in multistep avalanche chambers (section 2.5). A thorough review of ageing effects observed in the large wire chambers used in particle physics is given by Va'vra (1986). This paper discusses in detail the plasma chemistry involved in the creation of carbonaceous deposits on counter surfaces.

2.3.4   *Timing*

The intrinsic time resolution (in the sense of distinguishing between two events closely spaced in time) of a wire chamber is typically limited by positive ion mobility and anode–cathode spacing to the few microseconds level. The temporal data produced by a large-area proportional counter, however, usually owes more to the characteristics of the spacecraft clock, on-board computer and telemetry than to the gas physics of the detector.

A prime objective of the EXOSAT mission was to monitor regular and irregular X-ray intensity variations on timescales from $10\,\mu s$ to 80 hours (the useful observing time per orbit). This was achieved, in the case of the MEDA, through a number of versatile hardware and software modes (Turner *et al.*, 1981). On the shortest timescales, 'time-tagging' photon arrivals was controlled by the spacecraft clock frequency of 2^{17} Hz – a basic time digitisation of $7.63\,\mu s$. A variety of *high time resolution* (HTR) data handling modes provided broad-band temporal information on timescales up to 0.1 s. Thereafter, *high-energy resolution* (HER) modes of on-board computer operation accumulated eight, 32 or 128 channel energy spectra at the expense of timing information. Searches for X-ray pulsations could be automatically carried out by folding (Appendix B) the data at a frequency input in software. Some observations made with the MEDA are described in the next section.

The data modes designed for the Ginga LAC exhibit the same compromises between energy and timing information given the inevitable constraints on telemetry and on-board data storage (Table 2.3). The highest time resolution modes deliver the coarsest spectra (and least detector identification), while exploiting the full energy resolution of the counter can only be done with a time resolution of half a second.

2.3.5 *Observations*

Figs. 2.13–2.16, drawn from the three-year lifetime of the EXOSAT MEDA and from the first year of Ginga operations, illustrate the range of observations which can be made using a large-area, low-background proportional counter of modest energy resolution.

Fig. 2.13b is a MEDA spectrum of the supernova remnant W49B (Smith *et al.*, 1985a) which shows clear evidence of emission from highly ionised iron at 6.7 keV. Fig. 2.13a shows the best-fit model spectrum, corresponding to an optically thin plasma at a temperature of 2.6×10^7 K, with emission from the hydrogen and helium-like ions of Si, S, Ar, Ca and Fe (see Appendix B).

Figs. 2.14 and 2.15 illustrate the sensitivity of the MEDA to intensity variations on different timescales. The first of these figures shows the discovery of a 351 s modulation in the old nova GK Persei, corresponding to the rotation period of a compact white dwarf star in a binary system

Table 2.3. *Ginga Large Area Counter data modes*

Mode	Number of energy channels (0–35 keV)	Module identification[a]	Layer identification[b]	Time resolution[c] (ms)
MPC1	48	Individual modules	Yes	500
MPC2	48	Two groups	No	62.5
MPC3	12	None	No	7.8
PC	2	Two groups	No	0.98 or 1.9

[a] Identifies in which module (or group of four modules) the X-ray was absorbed.
[b] Identifies in which anode layer (R1, L1 or S23: fig. 2.12) the X-ray was absorbed.
[c] Values appropriate to the highest rates of data storage. The two figures in PC mode refer to the low- and high-energy channel, respectively.

(Watson *et al.*, 1985). Consistent determinations of the period were made using both Fourier power spectrum and epoch-folding techniques (Appendix B). Such long (~ 8 h) uninterrupted observations were a unique feature of the EXOSAT mission and its deep space orbit.

Fig. 2.15a shows a sequence of X-ray bursts from the source 2S1636–536 (Turner and Breedon, 1984) resulting from the 'flash' accretion of hydrogen on to the surface of a neutron star. The bursts are similar in profile and occur at regular intervals (Δt); the parameter α is the ratio of continuum luminosity to burst luminosity. The observed α values indicated that it is hydrogen, not helium, which is undergoing thermonuclear fusion at the surface of the compact object. Fig. 2.15b shows the discovery of a new phenomenon with the MEDA – intensity-dependent quasi-periodic oscillations (QPO) – first seen during a search for periodicities in the bright galactic source GX5-1 (van der Klis *et al.*, 1985). The broad peak in the power spectrum at ~ 20–40 Hz is believed to signify fluctuations in the accretion flow of matter on to the source. QPO have since been identified in nine more sources, including Sco X-1.

Fig. 2.13. (a) Best-fit model source spectrum for the supernova remnant W49B (Smith *et al.*, 1985). (b) EXOSAT MEDA spectrum for W49B (crosses) (in counts per second per detector module). The dotted line represents the prediction of the model spectrum after convolution with (blurring by) the MEDA's efficiency and energy resolution functions. χ^2 for this fit (Appendix B) was 53.7 for 46 degrees of freedom, implying significance at the $\sim 83\%$ confidence level. (*Courtesy A. Smith.*)

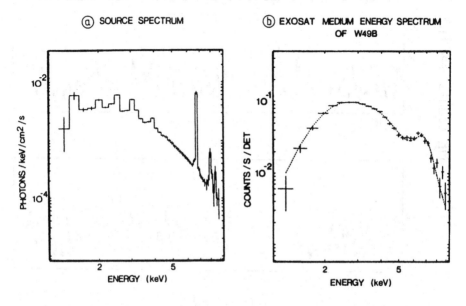

Fig. 2.16, finally, is the X-ray light curve of the SNR 1987A in the Large
Magellanic Cloud, measured by the Ginga Large Area Counter (Dotani
et al., 1987).

2.4   Imaging proportional counters

In imaging, as in non-imaging X-ray astronomy, gas-filled detectors, with their relatively large active areas, moderate energy and spatial
resolution and high sensitivity, commonly perform the 'workhorse' role.
Seventy per cent of all observations with the Einstein Observatory, for

Fig. 2.14. Background-subtracted 2.5–11 keV MEDA observations of the old
nova GK Persei, showing many cycles of the 351 s modulation (Watson *et al.*,
1985). The abscissa is in counts per second per half array. (*Courtesy M. G.
Watson and the Royal Astronomical Society.*)

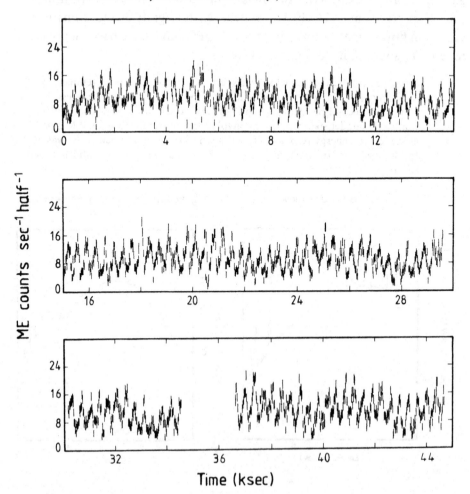

example, were made using the *imaging proportional counter* (IPC: Goren-stein *et al.*, 1981). This section describes the construction and operation of conventional (avalanche) imaging gas detectors with reference to three satellite-borne instruments:

(i) the Einstein IPC;

(ii) the EXOSAT parallel-plate avalanche chamber (commonly referred to as the PSD – the position-sensitive detector: Mason *et al.*, 1984);

(iii) the ROSAT position-sensitive proportional counter (PSPC: Pfef-fermann and Briel, 1985; Stephan and Englhauser, 1986).

All three of these detectors were used, or will be used, at the foci of grazing-incidence telescopes. A larger $(25 \times 25 \text{ cm}^2)$ MWPC, built at Utrecht for the coded aperture telescope on the Soviet Mir/Kvant space station, is described by Mels *et al.* (1988). Imaging gas scintillation and multistep counters are described separately, in section 2.5.

2.4.1 *Position encoding in proportional counters*

All methods of electronic position encoding lie somewhere between 'digital' and 'analogue' extremes. A digital encoder associates a preampli-fier and counting circuit with each resolution element ('pixel') of the field. An analogue system estimates event coordinates from the properties of voltage waveforms at (typically) four output electrodes. Digital schemes are, generally, capable of handling the higher count rates and are the more complex. In X-ray astronomy, where count rates are generally low and electronic simplicity is highly desirable, interpolative analogue readout, in which it is the centroid of the induced charge distribution that is encoded, is much the more commonly used.

Much of what follows is also relevant to imaging microchannel plate detectors (section 3.4), albeit on rather different position resolution scales.

2.4.1.1 *Signal location in one dimension*

The first position-sensitive proportional counters were cylindrical, single-wire devices (fig. 2.1a), developed as focal plane particle detectors for magnetic spectrographs, where they replaced nuclear emulsions (Ford, 1979). The first examples were described by Kuhlmann *et al.* (1966) and by McDicken (1967). In both these detectors the anode wire was used as a resistive divider and the fractional position of the avalanche along the wire,

x/L, inferred from the charges q_A, q_B received at the shorted ends A, B via the ratio:

$$Q_x = q_B/(q_A + q_B)$$

Ideally (fig. 2.17a):

$$Q_x = x/L \tag{2.13}$$

In practice, the amplitude ratio determined by an appropriate summing-and-dividing circuit (fig. 2.17c) deviates from this linear relationship away from the centre of the encoder as shown schematically in fig. 2.17b.

This charge division or amplitude ratio technique had first been developed for use in semiconductor detectors (Lauterjung *et al.*, 1963; Kalbitzer and Melzer, 1967). When applied to SWPCs, it produced position resolutions of about 1 mm in counters a few tens of centimetres long – fractional resolutions $\Delta x/L \simeq 2\text{–}4 \times 10^{-3}$. Non-linearities δ (rms deviations from the linear relationship eq. 2.13) of $\sim 1\%$ of the encoder

Fig. 2.15. (a) MEDA observations of X-ray bursts from source 2S1636-536 (Turner and Breedon, 1984). Count rates include the detector background of 10 counts/s. The break in the time-series at 12 000 s corresponds to an exchange of MEDA detector halves (section 2.3.1). (*Courtesy M. J. L. Turner and the Royal Astronomical Society.*) (b) Fourier power spectrum (Appendix B) of MEDA observations of the galactic source GX5-1 (van der Klis *et al.*, 1985) showing evidence of quasi-periodic oscillations – the broad peak between 20 and 40 Hz. (*Courtesy M. van der Klis. Reprinted by permission from* Nature, *vol. 316, p. 225.* © *1985 Macmillan Magazines Ltd.*)

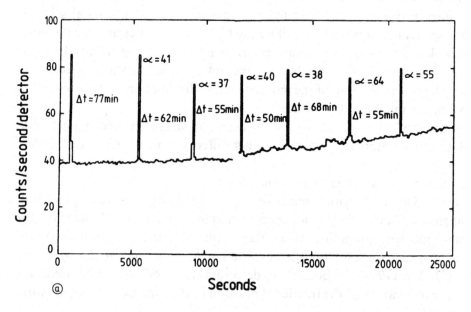

Fig. 2.15. (Continued)

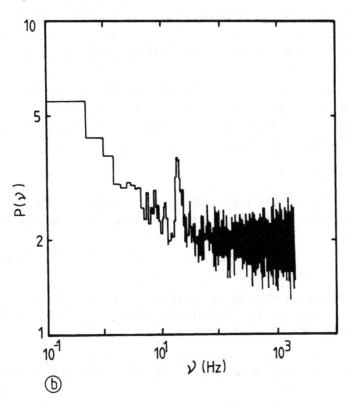

ⓑ

length were achieved. The anodes used were thin (~ 0.2 mm diameter) wires of low resistance.

A second method of signal location – the rise-time or zero-cross time method – was introduced by Borkowski and Kopp in 1968. By virtue of its capacitance to the cathode sheath, C, the central anode of a SWPC constitutes a distributed RC transmission line. The rise-time of the voltage pulse developed at an end of the line thus depends on the distance of charge injection from that end. In practice, both line-end pulses are shaped by doubly differentiating filters (fig. 2.17d) and the difference in zero-crossing times of the resulting bipolar waveforms is taken as a (reasonably) linear measure of the avalanche position. That is:

$$T_x = t_A - t_B \sim S_m(x/L - 0.5) \tag{2.14}$$

where S_m is the mean sensitivity, the gradient of the best-fit straight line approximation to T_x. Because the magnitude of S_m (and hence the image scale) depends on the product of the total anode resistance and its associated capacitance to ground, RC, and because the capacitance should be kept as low as possible in order to minimise electronic noise, SWPCs using the rise-time method typically have anode wires of much higher resistance than those using charge division. A common high-resistivity material for SWPC anodes is carbon-coated quartz fibre (10–80 kΩ/cm: Ford, 1979).

A comprehensive account of both amplitude ratio and rise-time encoding, based on an *RC* line model, is given by Fraser *et al.* (1981a, b). While the ratio method is ideally linear ($\delta = 0$) in a shorted encoder, in the limit of large normalised filter time constants (the so-called d.c. limit):

$$\pi^2 T_a / RC \gg 1$$

zero-cross timing is not inherently linear. Acceptable non-linearity ($\delta < 1\%$) can, however, be attained by careful choice of the normalised

Fig. 2.16. Ginga LAC 10–30 keV light curve of SNR 1987A (Dotani *et al.*, 1987). (*Courtesy K. A. Pounds. Reprinted by permission from* Nature, *vol. 330, p. 230.* © *1987 Macmillan Magazines Ltd.*)

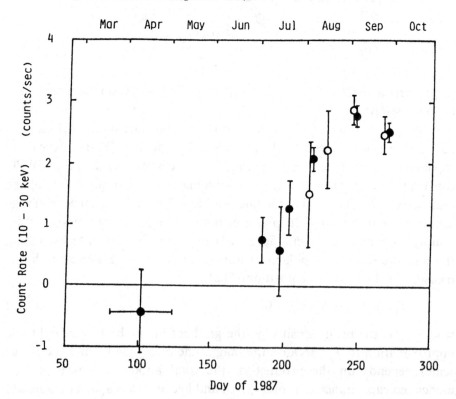

filter time constant and of the capacitances terminating the line. Regarding spatial resolution, it can be shown that zero-cross timing is, in principle, inferior to charge division by about a factor two. The rms encoder resolution (eq. 2.11) for these (and all other analogue encoding methods) has the form:

$$\Delta x_n/L = (\Delta x_a^2 + \Delta x_e^2)^{\frac{1}{2}}/L = D(q_a^2 + q_e^2)^{\frac{1}{2}}/q_0 = D(q_n/q_0) \tag{2.15}$$

Here, D is a dimensionless parameter (of order unity) whose functional form depends on the position signal being employed, q_0 is the signal charge collected by the anode and q_n is the result of adding in quadrature the rms equivalent noise charges of the encoder (suffix a) and of its terminal preamplifiers (suffix e). Equivalent noise charge for any particular noise source is that charge which, when dumped on the preamplifier input, would give the observed noise voltage.

Encoder limitations, however, are not always the most important factors in determining the performance of an X-ray imaging system. The precision, stability and dynamic range of the signal processing electronics (figs. 2.17b, d) must always be taken into account. The first X-ray resolutions measured with rise-time encoding ($\Delta x/L < 10^{-3}$: Borkowski and Kopp, 1968) were in fact as good as, or indeed rather better than, those achieved up to that time using charge division. Determining when a voltage waveform changes sign is rather easier than the analogue division of two signals. The 'state of the art' of position-sensing electronics in the early 1970s is reviewed by Lampton and Paresce (1974). Because of its electronic simplicity, the rise-time technique was rapidly applied in X-ray astronomy, both in one dimension, where the use of a position-sensitive SWPC with a curved diffracting crystal (fig. 2.18) allowed the registration of a range of X-ray wavelengths simultaneously, and in two dimensions as discussed below.

A third class of analogue encoder, which has certain advantages over those using timing or charge division, operates by charge sharing between a pattern of conductors. In SWPC design, these progressive geometry encoders are represented by the backgammon cathode of Allemand and Thomas (1976), shown unfolded onto a plane in fig. 2.17e. The fractional area occupied by one charge collector increases linearly from zero to unity over the length of the encoder. Solid angle calculations show that the amounts of avalanche charge received by the half-cathodes A, B (q_A, q_B, respectively) vary smoothly with avalanche coordinate x/L such that

$$(q_B - q_A)/(q_B + q_A) \simeq (x/L - 0.5)$$

A summing-and-dividing circuit similar to that shown in fig. 2.17c then provides a linear measure of avalanche position. Because the charge collecting electrodes are conductors, noise sources in this and other progressive geometry encoders are capacitive, rather than resistive as in *RC* lines. Progressive geometries are also rather faster (have a higher count rate capability) than *RC* lines for the same reason. The very partitioning of the cathode, however, gives rise to a new source of image blur. The arrival of the charge carriers (in this case ions) at the cathode segments constitutes a partition noise process (Martin *et al.*, 1981) governed by binomial statistics. Statistical variation in the division of the charge carriers between electrodes then blurs our estimate of the event position even in the absence of electronic noise. Position resolution due to partition noise varies as the signal charge q_0 to the power minus one-half, so that in high-gain applications (section 3.4.1) it may come to dominate over the more rapidly decreasing electronic noise contribution of eq (2.15).

2.4.1.2   *Signal location in two dimensions*

Following the (re)discovery that anode wires in a multi-wire structure could be made to act as independent counters (Charpak *et al.*, 1968) a large number of 2-d readout schemes were quickly developed to provide coverage of large ($>100 \, \text{cm}^2$) areas.

A 2-d readout system may consist either of a single two-dimensional encoder (both x and y coordinates from the same element) or two independent one-dimensional encoders arranged orthogonally. Proportional counter imaging further divides between anode-based methods and methods which utilise the charges induced on cathode planes. The

Fig. 2.17. Position encoding in SWPCs. (a) Signal location by resistive charge division. A current impulse $i_0 = q_0\delta(t)$ is injected to a uniformly distributed resistor, total resistance R, at coordinate x. The currents flowing into the shorted ends A, B are such that (i) $i_A + i_B = i_0$, and (ii) $i_A R_A = i_B R_B$. R_A and R_B are the resistances between the point of current injection and the ends of the resistor. Obviously: (iii) $R_A = (x/L)R$, and (iv) $R_B = (1 - x/L)R$. Manipulation of (i)–(iv) gives the result: $i_B/i_0 = i_B/(i_A + i_B) = x/L$. Integration over time then leads to equation (2.13). (b) Amplitude ratio Q_x. Broken curve = ideal. Full curve = measurement (the s-curvature is exaggerated for the purposes of illustration). (c) Schematic sum-and-divide circuit for realisation of amplitude ratio encoding. A, B represent the ends of the SWPC anode. (d) Schematic circuit for realisation of the rise-time method (Borkowski and Kopp, 1968; Parkes *et al.*, 1974). A, B represent the ends of the SWPC anode. The delay in channel A ensures that the pulses always arrive at the time-to-amplitude converter (TAC) in the same order. The magnitude of the TAC output is proportional to the difference in zero-crossing times $t_A - t_B$. T_a is the shaping filter time constant. (e) Division of the unfolded SWPC cathode in the scheme of Allemand and Thomas (1976).

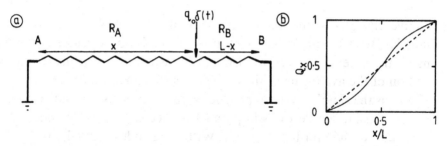

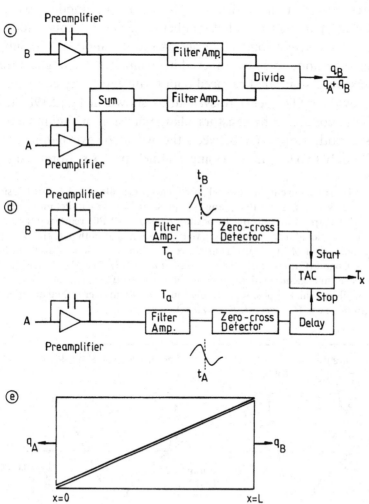

positive charge distribution on the cathode of a MWPC is essentially Gaussian, with a fwhm equal to about twice the anode–cathode separation and with a centroid (centre-of-gravity) displaced vertically from the position of the avalanche in the anode plane (Breskin *et al.*, 1977).

Perhaps the simplest way to produce area coverage is to sandwich many position-sensing anode wires in parallel between two cathode planes (fig. 2.19a; Borkowski and Kopp, 1975). With the anodes mutually isolated in this way, resolution in the y direction is set by the anode separation (~ 1 mm) for normal-incident radiation liberating charge in one anode cell only. That is, the image is 'binned' in one axis at the anode pitch. Simpler and cheaper 2-d readout may be achieved by completely 'analogue' means, at the expense of a somewhat reduced count rate capability compared to that of the two-amplifier-per-wire system, as shown in fig. 2.19b. If the anodes are interconnected by resistors, charge division or timing methods may be used both along and between the wires to give bidimensional readout with only two signal processing channels per axis. Borkowski and

Fig. 2.18. (a) Conventional crystal spectrometer (see also fig. 1.6a). (b) Using a curved crystal a range of Bragg angles is presented to the parallel input beam. A dispersed spectrum falls on the position-sensitive SWPC, removing the need for any rotational mechanism. A curved potassium acid phthalate (KAP) crystal spectrometer of this type was first used to examine the solar corona in the 10–18 Å band (Catura *et al.*, 1974; Rapley *et al.*, 1977). The SWPC detector, using rise-time encoding along a 50 µm diameter, 30 kΩ/cm carbon-coated quartz fibre, achieved positional resolutions $\Delta x \sim 1$ mm over a total encoder length of 18 cm. (*Courtesy J. L. Culhane.*)

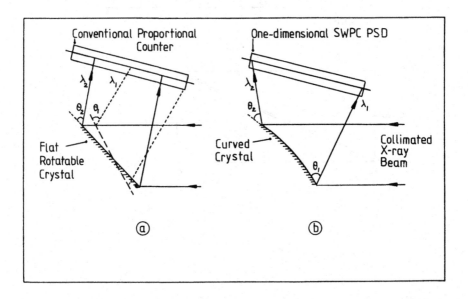

Fig. 2.19. Anode-based MWPC readout. (a) Area coverage by parallel *RC* line anodes. Two amplifier chains per anode. (b) Resistively interconnected anode wire plane. Two amplifiers per axis. Charge coordinate *x* is obtained from the difference in zero-cross times of the voltages X_A and X_B, when shaped by a bipolar filter. Charge coordinate *y*, similarly, may be derived from voltages Y_A and Y_B.

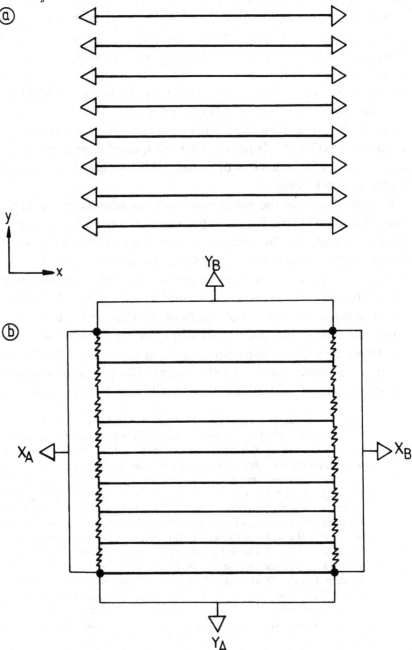

Kopp (1970) described a wire chamber of this type with 30 quartz fibre wires, each 12 cm long and separated from its neighbours by 2.5 mm. The chamber was used for 2-d imaging of α particles and low-energy (< 50 keV) X-rays. Normal-incidence resolution was limited to the wire spacing in y and to a few tenths of a millimetre in x.

With ionising radiation incident obliquely to such an anode plane, avalanches would obviously spread over several wires, and the centroiding property of the rise-time or ratio method would give a continuous measure of the event 'position' in y. The introduction of a lateral drift space to the counter (fig. 2.1c) provides the same interpolative effect for all angles of incidence by allowing the primary charge cloud to avalanche on to many wires. Even without such a drift space, angular localisation of the avalanche (section 2.2.2) provides an unexpected degree of interpolation in centroid-finding encoders operating with low gas gains (Gilvin et al., 1980b; Charpak, 1977).

Interconnection of the anode wires with resistors as shown in fig. 2.19b causes the x- and y-axis signals to become interdependent, leading to image distortion. A square object produces a 'pincushion' image (Borkowski and Kopp, 1975). In the limit of vanishingly small wire pitch, the distributed resistance of such a wire plane resembles that of a continuous resistive sheet. Continuous resistive anodes, produced by glazed thick film evaporation on ceramic substrates or otherwise (Curran, 1977), have in fact been widely used in parallel-plate avalanche chambers, where there is no need for the anode to be transparent to drifting ions. Fig. 2.20 illustrates the variety of possible resistive anode designs. The theory of resistive anode encoding is described by Fraser and Mathieson (1981a, b), who identify

Fig. 2.20. Uniform resistive anodes. The numbered electrodes are assumed connected through charge-sensitive preamplifiers to shaping filters of time constant T_a. (a) 1-d resistive strip, sensitive in the x direction only. (b) Square resistive anode with charge collecting electrodes central to its sides (Fraser and Mathieson, 1981a). Amplitude ratio position signals $Q_x = V_1/(V_1 + V_3)$; $Q_y = V_2/(V_2 + V_4)$, where V_j ($j = 1, .., 4$) are the peak electrode output voltages. Difference zero-cross time position signals $T_x = t_3 - t_1$; $T_y = t_4 - t_2$, where t_j ($j = 1, .., 4$) are the zero-cross times of the electrode output voltages. (c) Square resistive anode with electrodes at its vertices (Fraser and Mathieson, 1981a). Position signals as in (b). (d) Circular anode with point contact electrodes (Stumpel et al., 1973; Fraser and Mathieson, 1981b). Position signals as in (b). (e) Resistive anode of Gear's (1969) design. A sheet of uniform surface resistance $R_\square$ is bounded by circular arcs of equal radii a. Each arc should ideally be lined by a linear resistance $r_L = R_\square/a$: practical versions are bounded, as shown, by strip resistances of width w and surface resistance $r_\square = (w/a)R_\square$. Amplitude ratio position signals $Q_x = (V_1 + V_4)/(V_1 + V_2 + V_3 + V_4)$; $Q_y = (V_1 + V_2)/(V_1 + V_2 + V_3 + V_4)$.

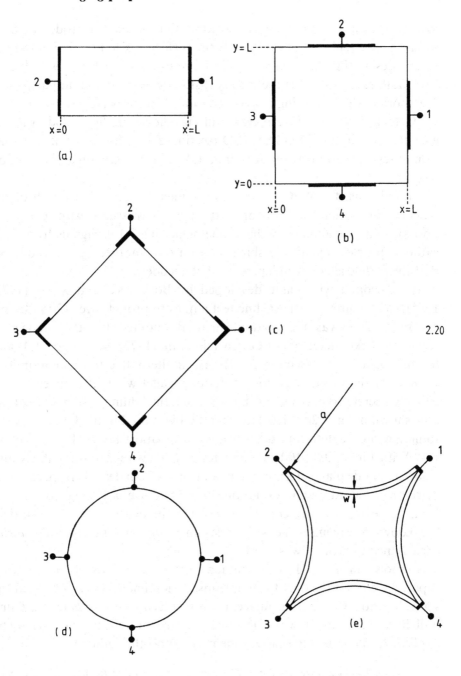

low distortion combinations of position signals and electrodes in both square and circular anodes of uniform surface resistance. The anode of Gear's design (fig. 2.20e: Gear, 1969; Lampton and Carlson, 1979; Fraser and Mathieson, 1981c) is the ideally distortionless 2-d counterpart of the 1-d shorted *RC* line using charge division. The resistive disc with point contact electrodes and zero-cross timing shown in fig. 2.20d was the encoder used in the EXOSAT PSD (section 2.4.3). More detailed discussion of resistive anode readout is undertaken in that section and in section 3.4.

Cathode-based MWPC readout is outlined in fig. 2.21. In each of the schemes illustrated, a pair of orthogonal ('crossed') wire planes is positioned above and below the anode plane (fig. 2.1c). Signals from each cathode are electrostatically shielded from one another by the anode, so that the cathodes act as independent 1-d encoders.

The Z-wound wire plane developed by Borkowski and Kopp (1972, 1975) was the first use of *RC* line techniques to provide area ($20 \times 20\,\text{cm}^2$) coverage. This was the readout method adopted for the pioneering sounding rocket detectors of Gorenstein *et al.* (1975) (see section 1.4) and for the Einstein IPC (section 2.4.2). As an alternative to this continuous cathode method, one may couple many parallel wires at their ends into either a discrete component *RC* line or into a thick film resistive strip of the type shown in fig. 2.20a. The latter method was used in the Leicester/MIT sounding rocket detectors described by Rappaport *et al.* (1979) (section 1.4). With 1 mm pitch, $100\,\mu\text{m}$ diameter cathode wires set 4 mm above and below an anode plane of 2 mm pitch, and with a 2 cm deep drift space, these detectors achieved spatial resolutions of $\sim 3\,\text{mm}$ over a 10 cm field. The gas gain, however, was set conservatively low in order to avoid electrical breakdown problems. The strip resistance was $50\,\text{K}\Omega$ and the *in situ* capacitance to ground was 30 pF.

As shown in fig. 2.21, a distributed resistance with its attendant stray capacitance is not the only form of transmission line which can be used for signal location. Connecting adjacent wires with capacitors creates a *CR* line (Gilvin *et al.*, 1980a), at whose terminals charge division algorithms may be applied. As a further alternative, one may couple the cathode wires into an

Fig. 2.21. Cathode-based MWPC readout. (a) Z-wound *RC* line for use with the rise-time method (Borkowski and Kopp, 1972, 1975). (b) Cathode wire plane coupled to an *RC* (wires connected by resistors) or *CR* (wires connected by capacitors) or *LC* transmission line. (c) Partitioned many-amplifier *RC* line readout. (d) Direct computation of the charge cloud centroid using cathode strips. Distribution of induced charge is indicated by relative magnitudes of preamp signals. All the encoders illustrated here are sensitive in the *y* direction

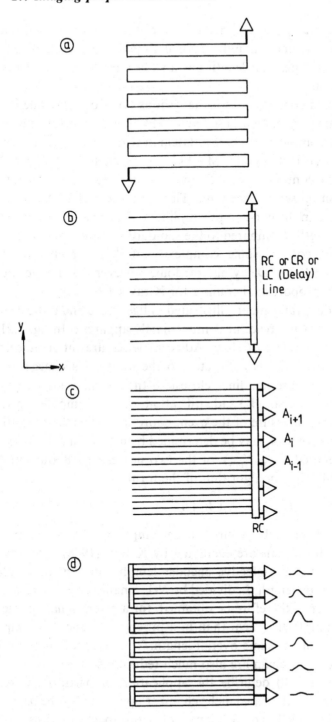

electromagnetic delay line and deduce the position of charge injection by observing the difference in pulse arrival times at the ends of the line, provided the delay per unit length has a sufficiently large (0.5–10 ns/mm) and uniform value.

The theory and construction of the various types of delay line (lumped parameter, coaxial, solenoidal) used in MWPCs for high-energy physics are described by Lecompte *et al.* (1978). Bateman (1984) describes the development of xenon-filled delay line MWPCs for fields such as digital X-ray radiography. In contrast with *RC* lines, delay lines are an ideally non-dissipative position-sensing medium. That is, noise arises only in the line terminations and in the line-end preamplifiers. High spatial resolution over large encoder lengths (compared to the standard *RC* line) may therefore be achieved at the cost of some complexity of design and manufacture (Lecompte *et al.*, 1978). Delay line readout, however, has not so far been reported in any proportional counter for X-ray astronomy.

In the cathode arrangements illustrated in figs. 2.21c and d the centroid-finding function of the readout is immediately apparent. In fig. 2.21c the encoder is a partitioned *RC* line. Adjacent wires are interconnected by resistors; every nth wire is connected to the input of a charge-sensitive preamplifier. The sense amplifier closest to the charge centroid (A_i, say) gives the largest signal (V_i). Identification of this amplifier gives an immediate 'coarse' measure of the event coordinate. The 'fine' position of the centroid relative to A_i may be determined by measurement of the signal amplitudes (denoted V_{i-1}, V_{i+1}), of the two nearest neighbour amplifiers (A_{i-1}, A_{i+1}) and using an algorithm of the form:

$$y/d = (V_{i+1} - V_{i-1})/(V_{i+1} + V_i + V_{i-1})$$

where d is the distance between adjacent amplifiers. More complicated, four-amplifier, algorithms are evaluated by Knapp (1978). For any such N-amplifier system, the effective encoder length entering the resolution eq. (2.15) is the total length, L, divided by $N-1$. Smaller absolute resolutions Δx (in microns) may therefore be obtained, for a given signal charge, in a 'hybrid' multi-amplifier system than in a purely analogue, two-amplifier-per-axis encoder of the same overall length. The demands placed on any one signal processing chain are also much reduced (section 3.4).

Bleeker *et al.* (1980) describe the application of partitioned *RC* line readout to a parallel-plate avalanche chamber optimised for the 0.05–1 keV XUV waveband. Orthogonal planes of close packed wires (100 μm diameter on 200 μm pitch) delineated the high-field region of the counter. Fwhm position resolutions in the range 120–250 μm were obtained in a 0.6 bar Ar–CH$_4$ gas mixture, albeit over a rather small imaging area ($1.6 \times 1.6 \, \text{cm}^2$).

Fig. 2.21d illustrates the direct computation of the centroid using cathode strips (Breskin *et al.*, 1977). If y_k ($k = 1, ..., N$) are the vertical locations of the preamplifier nodes, and V_k are the signal amplitudes at their respective outputs, the position of the avalanche along the y-axis is found by electronically realising the expression:

$$y = \sum_1^N y_k V_k / \sum_1^N V_k$$

One may either digitise the individual amplitudes using peak-sample-and-hold circuitry and analogue-to-digital converters (ADCs) and then compute the centroid in a microcomputer (Metzner, 1986) or, using convolution techniques, sequentially switch the strip outputs to a linear centroid-finding filter (Radeka and Boie, 1980). The 'centre-of-gravity' technique, originally developed for the fast localisation of minimum ionising particles (Breskin *et al.*, 1977) has been adopted for the ROSAT PSPC, described in section 2.4.4.

Returning once more to electronically less-complex two-amplifier-per-axis encoders, fig. 2.22 shows how area coverage may be achieved using the principle of charge sharing introduced in the previous section. Shown here for completeness is the four-quadrant anode (fig. 2.22a: Lampton and Malina, 1976; Wijnaendts van Resandt *et al.*, 1976; Mathieson, 1979; Purshke *et al.*, 1987) used in microchannel plate detectors to obtain very high resolutions ($\sim 10 \, \mu\text{m}$) over impracticably small field sizes.

The remaining progressive geometry encoders shown here have no such field limitations. A planar version of the backgammon anode of Allemand and Thomas (1976) is illustrated in fig. 2.22b. Fig. 2.22c shows, in one of its many forms, the wedge-and-strip (WS) anode (Martin *et al.*, 1981), an encoder which has been intensively developed in recent years for use in parallel-plate and multistep avalanche chambers, including the Penning Gas Imager (section 2.5), and for imaging MCP detectors (section 3.4.1). WS anodes are manufactured by photoreduction and etching techniques on Kapton or on glass substrates (Burton, 1982). In the three-electrode example shown, x-axis sensitivity is provided by variation in the thickness of the 'strips' and y-axis sensitivity by the wedges, just as in the backgammon anode. Obviously, charge must spread over several periods of the WS pattern to avoid periodic image non-linearities. A thorough discussion of the integral (δ) and differential (δ') non-linearities of WS encoding is given by Siegmund *et al.* (1986a). Differential non-linearity, the reciprocal of position sensitivity $\delta Q / \delta x$, measures the non-uniformity of response under uniform encoder illumination (in astronomical terms, the 'flat-field' properties of the detector).

Unlike the planar WS anode, the last of our encoding methods – the *graded density* (GD) grid (fig. 2.22d; Mathieson *et al.*, 1980) – is ideal for use in MWPCs. The GD grid is essentially a conventional cathode wire plane with the wires connected together in two groups, each of whose number densities varies linearly with position. Like the WS anode, GD grids may be produced 'in-house' in a suitably equipped research laboratory. Other encoder types, notably high-quality resistive anodes, generally cannot. Early versions of the GD encoder developed at Leicester showed appreciable differential non-linearity (Gordon *et al.*, 1983): δ' values comparable to those of uniform resistive anodes have now been demonstrated in improved designs. Two-fold subdivision of both WS (Siegmund *et al.*, 1986a) and GD (Gilvin *et al.*, 1981b) systems has been demonstrated in principle, but the improvement in spatial resolution which should accompany such division has not yet been measured in practice. Further consideration of the ultimate performance of both these and other readout methods is best reserved for the discussion of imaging microchannel plate detectors, where signal charges are generally much higher (10^7, rather than 10^5, electrons) and intrinsic detector limitations (eq. 2.11) are much less severe than in gas counters.

2.4.2   The Einstein Imaging Proportional Counter (IPC)

The scientific and technical characteristics of the highly successful Einstein IPC, built by American Science and Engineering (AS&E) from the sounding rocket detector design of Gorenstein *et al.* (1975), are summarised in Table 2.4. The IPC, together with its associated electronics and gas flow system, is shown in exploded view in fig. 2.23.

Fig. 2.22. Charge sharing (progressive geometry) encoders. Each numbered conductor is assumed connected to a charge-sensitive preamplifier and shaping filter. V_j ($j=1, .., 4$) are the peak electrode output voltages. (a) Four-quadrant anode. Quadrant conducting planes intercept the signal charge cloud (shaded). Position signals: $Q_x=(V_1+V_4-V_2-V_3)/(V_1+V_2+V_3+V_4)$; $Q_y=(V_1+V_2-V_3-V_4)/(V_1+V_2+V_3+V_4)$; or $Q_{x'}=V_1/(V_1+V_3)$; $Q_{y'}=V_2/(V_2+V_4)$. The latter method is inferior in terms of image distortion, yielding non-orthogonal coordinate estimates, i.e. $Q_{y'}$ depends on x'. (b) Planar backgammon anode of Allemand and Thomas (1976). Charge divides between the interleaved conductors. Position signal: $Q_x=V_1/(V_1+V_2)$. (c) Wedge-and-strip (WS) anode. One period of the WS pattern is shown. Electrode 1 = strip; electrode 2 (shaded) = zig-zag electrode; electrode 3 = wedge. Position signals: $Q_x=2V_1/(V_1+V_2+V_3)$; $Q_y=2V_3/(V_1+V_2+V_3)$. (d) Graded density (GD) wire plane, with wires connected in two groups. Position signal: $Q_x=V_1/(V_1+V_2)$.

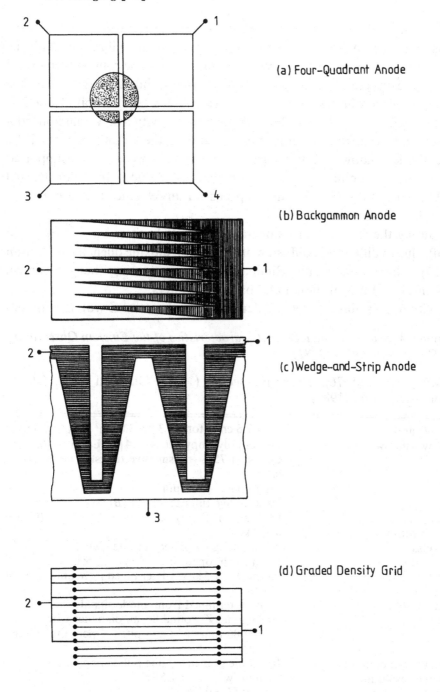

(a) Four-Quadrant Anode

(b) Backgammon Anode

(c) Wedge-and-Strip Anode

(d) Graded Density Grid

The purpose of the vacuum housing preceding the window of the IPC was to protect the thin plastic membrane from launch vibration. With the vacuum door open, a radioactive X-ray source could be commanded in and out of the detector field-of-view to provide an in-orbit gain calibration. The polypropylene window of the counter used exclusively in orbit (Table 2.4) was coated with Lexan in order to suppress sensitivity to UV emission from target stars and from the sun-lit earth. In both the redundant flight IPCs the plastic window and its tungsten support mesh were in turn supported by a set of aluminium 'ribs', whose shadows reduced the unobstructed field-of-view to $\sim 30' \times 30'$ and imposed an unwelcome structure on IPC images of diffuse sources.

During the first year of its operation, the IPC suffered from large gain drifts due to differential diffusion of the carbon dioxide quench gas (section 2.2.3). These were finally eliminated by reconfiguring the gas control systems of both operational and backup counters.

Within the stainless steel counter bodies, orthogonal Z-wound cathodes

Table 2.4. *Scientific and technical characteristics of the Einstein Observatory IPC (operational 1978–81)*

Harvey *et al.* (1976), Humphrey *et al.* (1978), Giacconi *et al.* (1979), Gorenstein *et al.* (1981).

Active area	7.6×7.6 cm² (total); 3.8×3.8 cm² (unobstructed)
X-ray window[a]	1.8 µm polypropylene + 0.4 µm Lexan + 0.2 µm carbon + 75% transmissive tungsten mesh (*A*), **or** 3 µm mylar + 0.2 µm carbon + 75% transmissive tungsten mesh (*B*)
Gas	84% argon + 6% xenon + 10% CO_2, 1.05 bar
Bandwidth (δE)	0.2–4 keV
Anodes	80 parallel gold-plated tungsten wires 1 mm pitch, operating voltage +3600 V
Energy resolution ($\Delta E/E$)	100% ($E > 1.5$ keV), determined from anode signal
Time resolution	63 µs, determined from anode signal
Cathodes	Orthogonal Z-wound 80 µm diameter wires looped 80 times at 1 mm pitch. Operating voltage +900 V
Encoding method	Rise-time plus delay line pulse shaping filter
Spatial resolution (Δx)	1.5 mm fwhm ($E > 1.5$ keV); 2.5 mm ($E = 0.28$ keV)
Maximum count rate	125 counts/s, telemetry limited

[a] There were, in fact, two IPCs on the Einstein focal plane 'carousel', identical but for their window composition. Only counter (*A*) was used in orbit.

(fig. 2.21a) recorded X-ray arrival positions in the Einstein focal plane. The Borkowski–Kopp method of encoding, chosen for its simplicity and low power consumption, did not, however, realise its full potential in the IPC. Relatively high noise levels in the preamplifiers and signal processing

Fig. 2.23. The Einstein Imaging Proportional Counter (IPC) (Harvey *et al.*, 1976). *Scale*: each counter grid is a square $10 \times 10 \, \text{cm}^2$ in area. (*Courtesy F. R. Harnden. © 1976 IEEE.*)

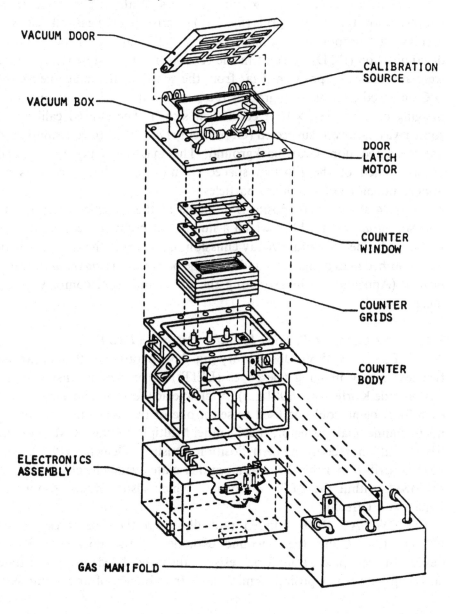

circuitry (described at the component level by Harvey *et al.*, 1976) limited the energy-dependent spatial resolution (eqs. 2.1, 2.15) to no better than 50 resolution elements per axis. Severe image non-linearity at the ends of each *RC* line also served to reduce the useful image area. Operation at high gas gains, resulting in degraded spectral resolution (compare the figures of Table 2.4 with the predictions of eq. 2.9), was a final consequence of the shortcomings of the image readout.

Nevertheless, the IPC provided X-ray astronomers with a powerful tool with which to observe faint, extended sources. With an anti-coincidence counter below the imaging grids (fig. 2.1c) subdividing the gas cell just as in a collimated proportional counter (section 2.3.2), and with the rise-time discrimination (RTD) inherent in the readout method (slow rising pulses were automatically positioned far from the centre of the image field), the IPC achieved in-orbit background rates comparable with previous non-imaging counters ($\sim 3 \times 10^{-3}$ counts/cm^2 s keV). The stowed calibration source was, however, suspected (Gorenstein *et al.*, 1981) of contributing to the measured non-X-ray background. The massive X-ray mirror and optical bench of the Einstein spacecraft must also have acted as an important source of secondary particles.

Fig. 2.24 shows a composite IPC image of the galactic centre region (Watson *et al.*, 1981a, b) in which a number of point sources are clearly visible embedded in diffuse X-ray emission. Comparing the raw data with the smoothed map of fig. 2.24b indicates the power of the *maximum entropy method* (Appendix B) in extracting the full detail from complex, noisy images.

2.4.3   *The EXOSAT Position Sensitive Detector (PSD)*

ESA's EXOSAT Observatory (fig. 2.6: operational 1983–6) carried two low-energy imaging telescopes (LEITs) as part of its instrumental payload (de Korte *et al.*, 1981a, b). A 'click-clack' exchange mechanism in each focal plane could be used to select one of two imaging detectors: a microchannel plate detector, the Channel Multiplier Array (CMA) or the (somewhat imprecisely named) Position Sensitive Detector (PSD). The PSD, described in Table 2.5, was a parallel-plate avalanche chamber built by MSSL within the LEIT consortium which also included groups at Leiden and Utrecht.

The EXOSAT PSDs differed radically from their predecessor, the Einstein IPC. Gone, the Z-wound cathodes of the Borkowski–Kopp design. In their place, a uniform resistive disc (fig. 2.20d) using rise-time encoding between opposing terminals. In the absence of anode-imposed binning, and with the higher proportional gas gains attainable, prior to the onset of space charge effects, in a parallel-plate geometry, the PSD promised, and achieved, much better spatial resolution (compare Tables 2.4 and 2.5) than a comparable MWPC.

X-rays entered the PSD gas cell through a double plastic window. The space between the polypropylene layers was vented to the skin of the spacecraft to prevent counting gas from the PSD reaching the nearby CMA, which required high vacuum for stable operation (section 3.2). The inner window, together with a nickel wire grid, defined a drift region of depth $d = 3.5$ mm (fig. 2.1e). Electrons from the initial ionisation then produced avalanches in the 1 mm (d') grid–anode gap, in fields of ~ 2 kV/ mm. The PSD resistive anode, operated at high positive potential, received signal charges of several picocoulombs per event at the higher X-ray energies, charge levels greater than in many microchannel plate detectors (section 3.3.2). Despite a strong dependence of gain on anode–grid spacing (Stumpel *et al.*, 1973, reported 1% gain changes for 1 μm changes in d' in an early laboratory PSD) energy resolutions close to the ideal of eq. (2.9) were achieved in the flight detectors, from analysis of the signals induced on the field-defining mesh. Background rejection was provided by the RTD inherent in the readout method and by use of four guard cells in an annulus surrounding the resistive disc. Also in this annulus was a gain monitor electrode permanently illuminated by 5.9 keV Mn K_α X-rays from an ^{55}Fe source. Signals from the monitor were used in an active gain control loop to compensate the grid–anode potential difference for changes in gas composition due to differential diffusion.

Clearly, the parallel-plate counter geometry has a number of advantages over any multi-wire arrangement; mechanical and electrical simplicity are foremost among them. The PSD's development history, however, reveals just how difficult it is to translate even a relatively simple laboratory concept into an instrument which works reliably in the unforgiving (and inaccessible) environment of space.

First, there were problems with the readout. The point-contact electrode, rise-time readout pioneered by Stumpel *et al.* (1973) was frozen into the EXOSAT design long before theoretical studies (Fraser *et al.*, 1981c) showed it to be inferior, in terms of distortion and resolution, to other resistive anode designs and long before other planar readout schemes – such as the WS anode – became available. The CERMET resistive film technology used in both CMA and PSD anodes (Curran, 1977) compounded the global 'barrel' distortion of the anode design with local non-uniformities which had to be carefully calibrated out of the final linearised images. Other resistive anode technologies (Barstow *et al.*, 1985) have since been shown to produce superior film uniformity and lower temperature coefficients of resistance than the CERMET technique. With the EXOSAT anodes, small variations in the focal plane temperature caused the detector image scale – determined by the anode time constant $R_\square C/\pi^2$ – to change significantly with time. Most seriously, however, the anode material became insulating when operated in an argon–carbon dioxide atmosphere (Cockshott and Mason, 1984). This precluded changing the Ar–CH$_4$ gas mixture, originally chosen to give optimum spatial resolution (Sanford *et al.*, 1979), when, during calibration of the LEITs two years before launch (Taylor, 1985), it was found that high-energy background events were producing localised sparks ('pings') in the PSDs. The resulting methane cracking (section 2.3.3) produced an insulating brown deposit on both the grid and anode defining the high-field region of the counter. Spurious low-charge pulses due to charging-up effects then followed. The sparks were of sufficient violence to vaporise small areas of the four gold electrodes on the resistive disc.

Spark-imposed constraints on counter lifetime were greatly alleviated by

Fig. 2.24. A 0.9–4 keV IPC image of the Galactic Centre region (Watson *et al.*, 1981a, b). (a) Raw data, binned in 32″ pixels. The image is the sum of two long (2.2 and 2.9 hour) observations made six months apart. (*Courtesy M. G. Watson.* © *1981 by D. Reidel Publishing company. Reprinted by permission of Kluwer Academic publishers.*) (b) Contour map after deconvolution and noise filtering by the maximum entropy method (Appendix B). The dynamical centre of our Galaxy (the radio source Sagittarius A West) is coincident with the intense X-ray source at 17h 42m 30s, −28° 59′ 01″. (*Courtesy M. G. Watson.*)

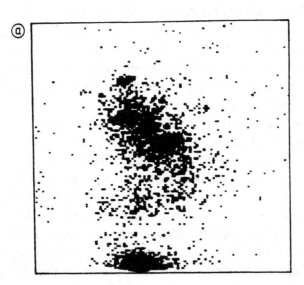

Fig. 2.24. (Continued)

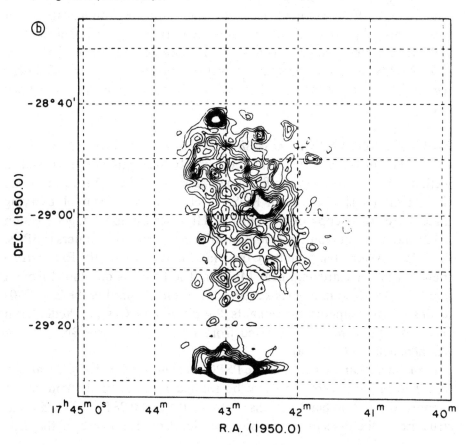

reducing detector capacitances and by the introduction of a small active device – the so-called 'ping-quencher' (Cockshott and Mason, 1984). This sensed the initial potential rise on the grid and rapidly reduced the avalanche field to zero, thereby suppressing the major part of each spark.

Successful though the ping-quencher was, both PSDs unfortunately suffered in orbit from problems reminiscent of those encountered in their ground development phase. The PSD of LEIT 2 failed completely after only 20 minutes of operation: that of LEIT 1 exhibited low-energy pulsing at its nominal voltage settings and had to be operated for the rest of the EXOSAT mission at reduced gain (and hence at reduced spatial resolution). The problem with PSD 1 may have been caused by a dust particle lodging close to the continuously illuminated gain control pad. The simple parallel-plate counter, with strong electric fields occupying a relatively large fractional volume of its gas cell, appears in hindsight a rather unforgiving device, perhaps too closely related to the spark counter (Fulbright, 1979) for comfort. The next generation of astronomical counter with a parallel-plate geometry, the multistep avalanche chamber (section 2.5), will certainly require much lower field strengths than the PSD. The susceptibility of these detectors, which retain large-volume avalanche regions, to particle-induced breakdown remains, however, an open question.

2.4.4 *The ROSAT Position Sensitive Proportional Counter (PSPC)*

The principal aim of the German ROSAT mission, described in section 1.5, is to perform the first imaging survey of the 0.1–2 keV X-ray sky. At the focus of the main ROSAT telescope is a 'carousel' carrying three detectors: a copy of the Einstein High Resolution Imager (section 3.4.2) and two redundant position-sensitive proportional counters (PSPCs: fig. 2.25) developed and built by the Max Planck Institut für Extraterrestriche Physik at Garching, near Munich. The PSPC is descended from a small (3×3 cm^2) sounding rocket detector originally used in the late 1970s to observe the supernova remnants Puppis A and Cas A (Pfeffermann *et al.*, 1981) and recently retrieved from storage to observe SNR 1987A (Aschenbach *et al.*, 1987a).

Pre-launch characteristics of the PSPC are described in Table 2.6 and in figs. 2.26a–f. A boron filter may be rotated in front of the counter to suppress the carbon band response between 0.1 and 0.28 keV. The internal structure of the detector is as shown in fig. 2.1c. The depth of the drift

Table 2.5. *Scientific and technical characteristics of the EXOSAT Observatory PSD*

Stumpel *et al.* (1973), Sanford *et al.* (1979), de Korte *et al.* (1981a, b), Cockshott and Mason (1984), Mason *et al.* (1984).

Active area	2.8 cm diameter (1.6° field-of-view)
X-ray window	Outer: 0.5 μm polypropylene + Lexan coating
	Inner: 0.8 μm polypropylene + 65% transmissive support mesh
Gas	80% argon + 20% methane, 1.1 bar
Bandwidth (δE)	0.1–2 keV
Image readout	CERMET resistive disc, rise-time encoding
Spatial resolution	0.2 mm fwhm ($E=1.5$ keV);
(Δx)	1.1 mm fwhm ($E=0.28$ keV)
Energy resolution ($\Delta E/E$)	$42/E^{\frac{1}{2}}$%, determined from grid signal
Background rate (B_i)	0.7 counts/cm^2 s keV

region (d) is 8 mm. Three imaging wire planes (C1, A, C2) share the gas cell with a two-plane (ACO1, ACO2) anti-coincidence counter. The imaging planes have been carefully redesigned to give maximum immunity to the damaging particle-induced sparks which were a problem early in the counter development phase (Pfeffermann and Briel, 1985). Accurate spacing of the anode wires and the use of a many-amplifier (25 per image axis) centroid-finding readout (fig. 2.21d) have produced, in the PSPC, full-field energy resolutions and image uniformities much superior to those of the similarly sized Einstein IPC (compare Tables 2.4 and 2.6). The spatial resolution of the PSPC in orbit will be somewhat degraded by X-ray penetration effects (section 2.2.5: the cone angle of the ROSAT X-ray telescope varies from 5.5–9°) to around 0.35 mm fwhm at 1 keV.

Such imaging performance cannot, of course, be achieved without cost in

Fig. 2.25. The ROSAT PSPC, its front cover removed, showing (i) mounting of wire grids on Macor (machineable glass ceramic) frames, and (ii) electrical feedthroughs (Briel and Pfeffermann, 1986). (*Courtesy E. Pfeffermann.*)

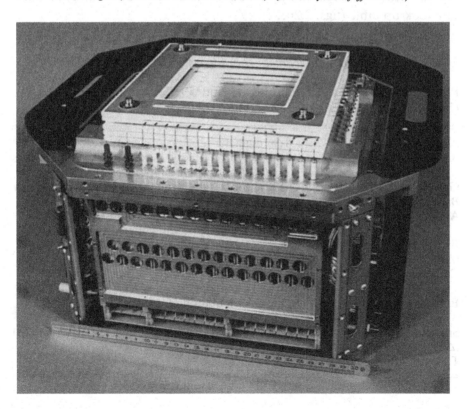

terms of experiment mass, power and electronic complexity. In this last area, especially, the PSPC represents a radical departure from all previous satellite-borne proportional counters.

Metzner (1986) describes the 'front end' (FE) signal processing electronics for the PSPC. A single FE unit handles signals from either operational PSPC, through a mixer unit. The analogue peak heights from all the charge-sensitive preamplifiers (56 per PSPC module) are first digitised and event selection logic is applied. Energy selection (section 2.3.2) is performed on the anode signal and a five-sided coincidence logic is applied to the signals of the monitor counter (ACO1, ACO2) and to the outermost 'guard' strips of both cathode planes. In addition, a form of pattern recognition – preselection logic (PSL) – is applied to the cathode signatures. For an event to be valid as an X-ray signal: (i) all the excited cathode strips must be adjacent, and (ii) the number of excited strips must not be greater than five. The additional sensitivity to charged particle tracks provided by PSL increases the PSPC ^{60}Co rejection efficiency to 99.5%. Valid event signals are passed to a 16-bit microprocessor (Motorola 68000) for computation of the charge centroid coordinates x, y. Most of the image non-uniformity inherent in the readout is finally calibrated out by recourse to a look-up table (figs. 2.26e, f).

Table 2.6. *Scientific and technical characteristics of the ROSAT PSPC*

Pfeffermann and Briel (1985), Briel and Pfeffermann (1986), Metzner (1986), Stephan and Englhauser (1986), Pfeffermann *et al.* (1987).

Active area	8 cm diameter (2° field-of-view)
X-ray window	1 μm polypropylene + Lexan and carbon coatings + 70% transmissive support mesh
Gas	65% argon + 20% xenon + 15% methane, 1.5 bar
Bandwidth (δE)	0.1–2 keV
Anodes	10 μm diameter gold-plated tungsten wires, 1.5 mm pitch, operating voltage +3100 V
Energy resolution ($\Delta E/E$)	$41/E^{\frac{1}{2}}$%, determined from anode signal
Cathodes	50 μm diameter platinum–iridium wires, 0.5 mm pitch, operating voltage +300 V
Encoding method	Direct computation of centroid from 25 cathode strips, each of seven wires
Spatial resolution (Δx)	0.25 mm fwhm ($E = 0.93$ keV, parallel illumination)
Time resolution	120 μs

The FE consumes 21 W of electrical power and accounts for some 30 kg of the total PSPC mass (excluding the gas replenishment system) of ~ 50 kg.

2.4.5 *Applications*

X-ray imaging proportional counters are being increasingly used in fields as diverse as non-destructive evaluation (MacCuaig *et al.*, 1986) and tumour location (Webb *et al.*, 1984). It is, however, rather difficult to pinpoint direct links between the astronomical counter technologies of the three previous sections and any ground-based application. The reason is mainly one of bandwidth.

The imaging PCs described above have all been small (<10 cm diameter), thin-windowed, low-pressure (<1.5 bar) argon-based devices, designed to operate only up to the (few kiloelectronvolts) cutoff energy of their accompanying X-ray optics. Ground-based X-ray imaging or tomography typically requires much higher energies: 60 keV and 511 keV, respectively, in the two examples cited above. Even for the 8 keV Cu K X-rays commonly used in diffraction studies of muscle (Faruqi, 1975; Faruqi *et al.*, 1986), the detection efficiency of the ROSAT PSPC (say) would only be around 40%.

More closely aligned with ground-based needs are the large (30×30 cm²) sealed, high-pressure, xenon-based counters developed for astronomical coded aperture imaging (Thomas and Turner, 1983, 1984; Willmore *et al.*, 1984; Mels *et al.*, 1988). Some of these MWPCs, using the readout methods described above (GD cathodes, centroid computation), have achieved spatial resolutions of ~ 1 mm fwhm over fields of 30 cm. The work of Thomas and Turner (1983, 1984), who describe, in an astronomical context, the physical limitations to spatial resolution and energy resolution in large xenon-filled counters, should be of great interest to many experimenters.

2.5 **Proportional counters with enhanced energy resolution**

The very first gas scintillation proportional counters (GSPCs), developed at the University of Coimbra, in Portugal (Policarpo *et al.*, 1972, 1974) achieved an energy resolution

$$\Delta E/E = 0.20/E^{\frac{1}{2}}$$

which, at a factor of two better than the equivalent figure of merit in conventional (avalanche) counters (sections 2.2.2, 2.2.4), immediately brought the device to the attention of the X-ray astronomy community.

Within a decade, a number of GSPC spectrometers had been flown on sounding rockets and satellite platforms (sections 2.5.1 and 4.3) and considerable progress, using a variety of photomultipliers, had been made in the area of imaging GSPCs (section 2.5.2). The main foci of GSPC research have been within the Space Science Department (SSD) of ESTEC (ESA's European Space Research and Technology Centre in the Netherlands), at Columbia University in the USA, and at ISAS in Japan.

As research into imaging GSPCs has matured, however, a spatial resolution limit, irrespective of photomultiplier design, of about 1 mm fwhm at 1 keV has become apparent. For this reason, and because the pure noble gases used in GSPCs are extremely intolerant of impurities, interest has grown in ways of avoiding the energy resolution constraints imposed by avalanche fluctuations in conventional counters, while retaining the submillimetre spatial resolution and relatively lax cleanliness requirements of the MWPC. The first of these new detectors used the principle of electron counting (Siegmund *et al.*, 1982), whereby the number of individual light pulses from primary electrons drifting in an argon-based gas mixture is used as an index of the initial X-ray energy. Electron counting is described in section 2.5.3. Finally, section 2.5.4 describes the Penning Gas Imager (Schwarz and Mason, 1984) and other multistep avalanche chambers (fig. 2.1f) which currently hold out most promise of combining in one gas cell the desirable properties of both GSPC and MWPC.

2.5.1 *Non-imaging GSPCs*

2.5.1.1 *Light collection*

The physical principles of the GSPC have already been described in sections 2.2.2 and 2.2.4. Obviously, the key to a successful detector is the

Fig. 2.26. ROSAT PSPC ground calibration data (Pfeffermann and Briel, 1985; Briel and Pfeffermann, 1986; Pfeffermann *et al.*, 1987). (a) Energy resolution for Cu L (0.93 keV) X-rays as a function of position perpendicular to the anode wires. (b) Pulse magnitude P (section 2.2.2) measured as a function of position perpendicular to the anode wires. (c) Quantum detection efficiency $Q(E)$. Note the narrow band of window transmission below the carbon edge in energy. Above 1.5 keV, Q is equal to the window transmission, T_w. (d) Fwhm position resolution, measured as a function of position perpendicular to the anode wires. (e) Image non-linearity. Displacements between true and encoded X-ray illuminated pinhole positions. The largest excursions are caused by field non-uniformities due to bulging of the counter window. Other, smaller, effects include digitisation due to discrete anode wires and charge thresholds in cathode strip readout. (f) As (e), except that observed positions have now been corrected according to a previously determined 'look-up' table. (*Courtesy E. Pfeffermann.*)

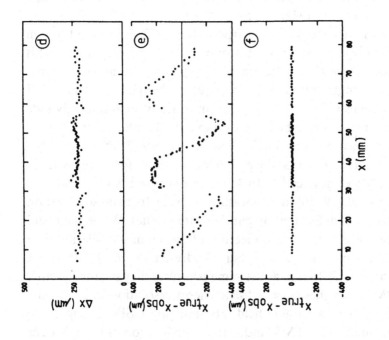

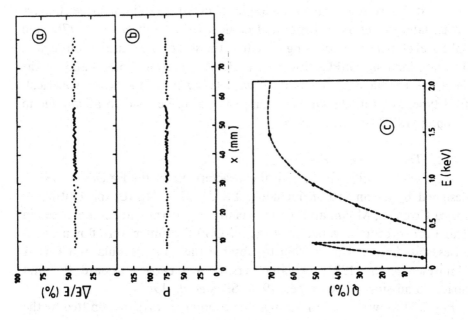

efficient production and collection of the 1500–1950 Å UV light which results from the various atomic and molecular excitation and de-excitation processes in the xenon gas (Manzo *et al.*, 1980).

In the pioneering detectors of Policarpo *et al.* (1972, 1974) scintillation occurred in the spherical field region around a ball-and-rod anode. This region was viewed through UV transmissive (Spectrosil quartz) windows, coated with a wavelength shifting layer of p-quaterphenyl, by a pair of EMI photomultiplier tubes (PMTs). The operation of traditional multi-dynode photomultiplier tubes is described, for example, by Knoll (1979).

The X-ray entrance window in the first laboratory GSPC was rather small, 12.5 cm^2 in area. The first development detectors for X-ray astronomy were much larger devices (up to 11 cm in diameter: Anderson *et al.*, 1977a; Andresen *et al.*, 1977b) incorporating separate drift and scintillation regions (as in fig. 2.1e) delineated by parallel wire meshes. While such field geometries were useful for the identification of optimum scintillation field strengths (~ 3250 V/cm in Xe at 1 bar; Andresen *et al.*, 1977b: light production starts at ~ 500 V/cm at the same pressure) they resulted, with a single output PMT, in rather poor full-field energy resolutions. Moving the X-ray source off-axis in a parallel-field, non-imaging GSPC changed the solid angle subtended to the PMT and, hence, the fraction of the UV light collected. To overcome this solid angle effect while retaining large effective areas, later detectors incorporated conical (Anderson *et al.*, 1977b) and spherical electrostatic focusing to reduce the size of the scintillation region. The non-imaging GSPCs flown on EXOSAT (section 2.5.1.2) and on the Japanese Tenma satellite (Inoue *et al.*, 1982) were of the latter, spherical-field type. Solid angle variation can, of course, be used to advantage in imaging GSPCs (section 2.5.2).

2.5.1.2   *The EXOSAT GSPC*

The scientific and technical characteristics of the EXOSAT GSPC, designed by a consortium including ESA's SSD, MSSL, the Istituto di Fisica Cosmica, Milan, and the University of Palermo, are summarised in Table 2.7. Prototype versions of the EXOSAT counter were flown on an Aries sounding rocket in 1980 to observe the Crab Nebula and Cas A (Andresen *et al.*, 1981). A later version was carried aboard the first Spacelab mission in December, 1983 (Sims *et al.*, 1985).

Fig. 2.27 shows a section through the counter. X-rays transmitted by the GSPC's microchannel plate collimator blocks (not shown) entered the drift region of the counter via a spherical section, free-standing beryllium window, which limited the low-energy response to about 2 keV. The counter body was constructed to ultra-high-vacuum standards from a number of machinable ceramic sections and baked to a temperature of 300°C to minimise contamination of the pure noble gas filling. For the same reason, wavelength shifting coatings were omitted from the UV exit window. Purity of the sealed counter volume was maintained by a pair of getter pumps. Six gold-plated molybdenum focusing rings on the inside of the ceramic cone established a spherical electric field, leading electrons into a scintillation region defined by two curved molybdenum grids.

The gas filling was a mixture of xenon and helium. The small helium component was necessary in order to improve background rejection by increasing electron drift velocities from the very low values observed in xenon (Inoue *et al.*, 1982). The temporal width of the UV burst length distribution arising from pointlike X-ray absorption, which must be as narrow as possible in order to successfully employ *burst length discrimination* (BLD: Andresen *et al.*, 1977a), is a strong function of drift velocity, particularly of that in the photon absorption region. In orbit, particle rejection by BLD was 80% efficient at the lowest X-ray energies. Overall rejection efficiencies (BLD plus energy 'windowing': section 2.3.2) exceeded 85% over the complete bandpass of the detector (Peacock *et al.*, 1985) and approached 97% at 2 keV. The residual background spectrum included line features at 10.54 keV (lead Lα fluorescence from the glass collimator) and 12.7 keV (a blend of lead Lβ fluorescence and thorium Lα emission, arising from plutonium contamination of the beryllium window) which, after considerable analysis and some misidentification, were used for in-orbit calibration of the detector's energy scale.

Table 2.7. *Scientific and technical characteristics of the EXOSAT GSPC*

Peacock *et al.* (1980), (1981), (1985).

Geometric area (A_s)	310 cm^2
Effective area (QA_s)	~ 100 cm^2
X-ray entrance window	175 µm beryllium^a
UV exit window	4 mm Suprasil quartz
Gas	95% xenon + 5% helium, 1.0 bar
Bandwidth (δE)	2–18 keV (2–40 keV, depending on PMT gain)
Full-field $\Delta E/E$	$27/E^{\frac{1}{2}}$% fwhm
Internal background after rejection (B_i)	1.3 counts/s keV (2–10 keV)
Collimator transmission	45′ fwhm (3′ flat-top)

^{a} The dome shape of the window means that the projected absorbing thickness increases towards the edge of the field-of-view.

Post launch, it was discovered that each X-ray spectrum produced by the GSPC contained a line at 4.78 keV. This subtle artefact was traced to a discontinuous change in the final ionisation state of, and hence the amount of energy retained by, the xenon atoms for X-ray energies above and below the xenon L III edge (Peacock *et al.*, 1985). The assumption contained in eq. (2.1), that w, the average energy required to create a secondary ion pair, is independent of the initial X-ray energy, is thus not wholly true.

The merits of the EXOSAT GSPC as a spectrometer can be judged by comparing its energy resolution with that of the EXOSAT MEDA (Tables 2.2, 2.7). The factor of two improvement provided by the GSPC has led to a number of important spectoscopic discoveries. Figs. 2.28a and b illustrate the large changes in the apparent energy of an iron emission line in the

Fig. 2.27. The EXOSAT Gas Scintillation Proportional Counter (Peacock *et al.*, 1980, 1981, 1985). (*Courtesy A. Peacock © 1981 by D. Reidel Publishing company. Reprinted by permission of Kluwer Academic publishers.*)

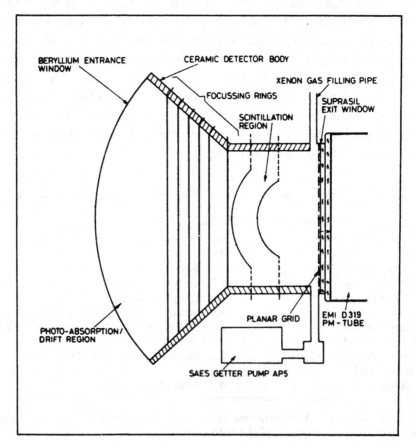

Fig. 2.28. EXOSAT GSPC X-ray spectra of SS433 (Watson *et al.*, 1986). The histograms are the best-fit spectra consisting of a two-component continuum plus an emission line (rest energy 6.7 keV). (a) Observation made at phase 0.31 of the 164-day precession period. (b) Observation made at phase 0.64 (*Courtesy M. G. Watson and the Royal Astronomical Society.*)

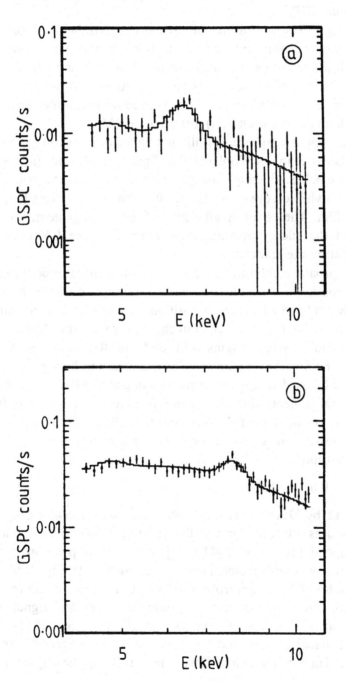

galactic source SS433, as a function of the precession phase of the source's relativistic jets. These changes are consistent with the Doppler shifts predicted by a kinematic model of the jets (Watson *et al.*, 1986).

2.5.2   *Imaging GSPCs*

Imaging GSPCs, utilising, of necessity, a parallel-field geometry (fig. 2.1e) divide into four categories, according to the type of position-sensitive readout used to localise the burst of UV light which follows X-ray absorption. These readout methods are: (i) Anger Camera arrays of photomultiplier tubes (PMTs); (ii) UV sensitised microchannel plates; (iii) photoionisation chambers; and (iv) multi-anode PMTs.

The first attempt to image the scintillation light from a xenon gas cell was reported by Gorenstein and Topka (1977). Spatial resolutions (at 5.9 keV) of 0.5 mm were achieved using Cassegrain optics to focus the xenon light on to a UV-to-visible converter, hence onto an image intensifier, and, finally, onto film. The counting efficiency of this arrangement, however, was low, and the energy information contained in the UV burst was completely lost by the readout.

The first approach combining imaging with good energy resolution was to subdivide the UV exit window between a number of PMTs, arranged symmetrically, and deduce the event position from the relative magnitudes of their outputs. This is the principle of the Anger Camera (Anger, 1958) developed for clinical γ-ray imaging with solid scintillators such as NaI(Tl) (section 5.2). The first Anger Camera GSPC appears to have been that of Hoan (1978), who used analogue centroid-finding algorithms (fig. 2.29a) in a five-PMT arrangement, obtaining spatial resolutions of ~ 2.7 mm fwhm over a field 11 cm in diameter. This counter, filled with xenon at one atmosphere pressure, was proportional over the energy range 2–35 keV with energy resolution

$$\Delta E/E = 0.01 + 0.22/E^{\frac{1}{2}}$$

As pointed out by Davelaar *et al.* (1980), such simple centre-of-gravity readout introduces substantial image distortion because of the complicated solid angle dependence of each PMT's response on event position. A more subtle, though more cumbersome, approach is illustrated in fig. 2.29b. The polar coordinates of X-ray absorption, relative to the detector centre, may be numerically determined from appropriate ratios of PMT signals once the variation in solid angle subtended by each PMT at the event position has been calculated. Using such a 'look-up table' approach, spatial resolution of 2.5 mm fwhm (at 6 keV) was measured with the original three-

Fig. 2.29. Arrangements of PMTs for readout of imaging GSPCs. The exit aperture of the UV light producing region is indicated by the broken circle. Let the PMT output signal amplitudes be P_j ($j = 1, .., 5$). Then: (a) in the centre-of-gravity method of Hoan (1978), X-ray coordinates are computed from $x = (P_3 - P_1)/(P_1 + P_3 + P_5)$; $Y = (P_2 - P_4)/(P_2 + P_4 + P_5)$, while the energy of the X-ray is computed from the sum of P_j. (b) In the PMT arrangement of Davelaar *et al.* (1980), the ratios P_j/P_{j+1} and P_j/P_{j+2} (defined cyclically) uniquely determine the X-ray position when compared with a theoretical look-up table. Image reconstruction by maximising an objective function (akin to the entropy) of the PMT count rates is described by Manzo *et al.* (1985).

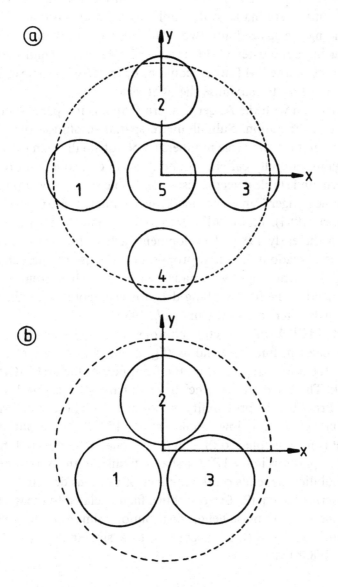

PMT arrangement (Davelaar *et al.*, 1980) in a detector 7 cm in diameter. Later work at SSD using a seven-PMT Anger Camera (Davelaar *et al.*, 1982) resulted in spatial resolutions (where *E*, as usual, is in keV):

$$\Delta x = 4.1/E^{\frac{1}{2}}\, \text{mm fwhm}$$

(1.7 mm at 6 keV) over active areas up to 12 cm in diameter. Energy resolution in these devices was maintained at a high level, viz:

$$\Delta E/E = 0.019 + 0.196/E^{\frac{1}{2}}$$

over an energy band extended below 1 keV by the use of thin (3.5 μm) mylar windows. Attempts were made to decouple energy and position determination by dividing the gas cell into two drift and two scintillation regions, in effect producing a multistep GSPC (section 2.5.4). The second scintillation region, closest to the PMT array, could be operated with some electron multiplication in order to maximise the light yield.

The main problem with the Anger Camera GSPC is its sparse sampling of the UV light distribution. Submillimetre spatial resolution over large fields-of-view would require large numbers (> 50) of small (1 cm diameter) PMTs, with gains carefully calibrated and actively controlled in order to ensure a uniform and stable detector response. While such a development is not inconceivable (Anger Cameras in medicine routinely use arrays of up to 75 PMTs: Short, 1984), multi-PMT detectors for astronomy rather fell from favour in the early 1980s. Development instead focused on 'continuous' methods of readout. The first proposal for the use of microchannel plates to image the xenon light was made by the Columbia group in 1981 (Hailey *et al.*, 1981). The first working detector was reported by the SSD researchers shortly afterwards (Taylor *et al.*, 1983).

In the GSPC/MCP X-ray detector, photons reach the evacuated MCP cell through a calcium fluoride window (fig. 2.30a) and release electrons from a CsI photocathode deposited on the front face of the first MCP of a two-plate stack. The UV quantum efficiency of a bare MCP is very low (see section 3.3.3). Provided the probability of two or more photons entering a single microchannel remains low (Hailey *et al.*, 1981), the output signal from the MCP is proportional to the number of channels activated, hence to the number of photons in the UV burst and, finally, to the X-ray energy. The energy resolution depends on the number of photons detected, N_{p}, in the manner described in eq. (2.10) (to within a factor related to the width of the MCP's single-electron pulse height distribution: Simons *et al.*, 1985b). Spatial resolution varies with N_{p} according to a similar $N_{\text{p}}^{-\frac{1}{2}}$ ($E^{-\frac{1}{2}}$) law (Simons *et al.*, 1985b) viz:

$$\Delta x = 2.36 \, (\sigma_0^2/N_p + \sigma_{xy}^2/N)^{\frac{1}{2}} \qquad (2.16)$$

where σ_0 is the standard deviation of the 1-d distribution of activated microchannels and σ_{xy} is the rms width of the primary electron cloud leaving the scintillation region (eq. 2.4). N is the number of primary

Fig. 2.30. Imaging gas scintillation proportional counters. (a). Cross-section through microchannel plate GSPC (Sims *et al.*, 1984). The distances z_1 (UV photocathode to bottom of scintillation region) and z_0 (photocathode to top of scintillation region) must be optimised for maximum efficiency of light collection (Taylor *et al.*, 1983). (b) Cross-section through photoionisation chamber GSPC (Sims *et al.*, 1984 after Hailey *et al.*, 1983). The X-ray window material is 1 micron polypropylene, coated with carbon and aluminium. Here, IPC denotes 'imaging proportional counter'. (*Courtesy M. R. Sims.*)

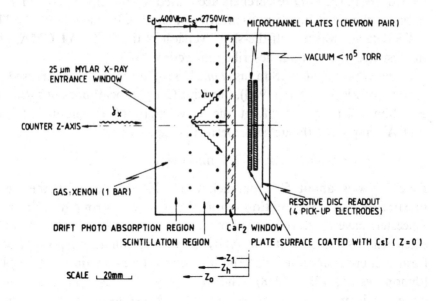

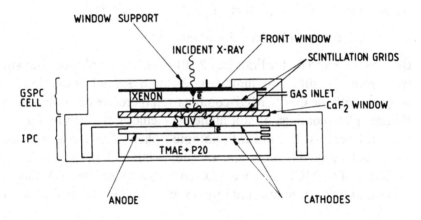

photoelectrons (eq. 2.1). Although the number of UV photons generated per X-ray event can be very large ($\sim 10^3$ per steradian per millimetre scintillation field), it is still essential that the 1500–1900 Å quantum detection efficiency of the MCP and its photocathode, averaged over all angles and positions of UV incidence, exceed $\sim 10\%$ if the predicted performance of the GSPC/MCP combination –

$$\Delta E/E = 0.20/E^{\frac{1}{2}}; \ \Delta x = 1.22/E^{\frac{1}{2}} \ \text{mm fwhm}$$

– is to be realised in practice.

The initial results reported by Taylor *et al.* (1983) were higher, by a factor three in energy resolution and a factor ten in spatial resolution, than these limits. While claiming an adequate quantum efficiency for their CsI coated MCP, the SSD researchers accounted for this poor performance in terms of the gross image non-linearity of the MCP image readout. Their MCP detector was an 'off-the-shelf' version of the EXOSAT CMA, using the resistive disc readout described in section 2.4.3.

Later work at Leiden (Simons *et al.*, 1985b), using the linear crossed-grid encoder of Bleeker *et al.* (1980), but with CsI photocathodes of rather low quantum efficiency ($<5\%$ at the wavelength of maximum emission, 1700 Å), improved the detector performance to a level:

$$\Delta E/E = 0.36/E^{\frac{1}{2}}; \ \Delta x = 4.2/E^{\frac{1}{2}} \ \text{mm fwhm}$$

Here, N_p was about 500 for each 6 keV X-ray. Recently, the Leiden researchers, in collaboration with the X-ray Astronomy Group at Leicester, have produced CsI photocathodes with rather better quantum efficiencies ($\sim 9\%$ peak at 1700 Å) by carefully eliminating exposures of the fresh vacuum-evaporated material to any moisture containing atmosphere (Simons *et al.*, 1987, 1988). The preparation of CsI photocathodes is discussed in more detail in section 3.3.3. These new efficiencies should translate into a GSPC performance characterised by:

$$\Delta E/E = 0.19/E^{\frac{1}{2}}; \ \Delta x = 2.1/E^{\frac{1}{2}} \ \text{mm fwhm}$$

close to the Fano factor limit (eq. 2.10) in energy resolution, but still some way from the diffusion limit (eq. 2.16) in spatial resolution. Further progress with the GSPC/MCP combination therefore depends on the use of channel plate coatings of still higher UV efficiency, such as Cs_2Te or the bialkali layers used in optical photomultipliers. These coatings are, however, much more difficult to prepare and maintain than CsI (Somner, 1968). Problems of (i) MCP gain degradation with accumulated UV flux, and (ii) gain suppression at high count rates (Sims *et al.*, 1984) may further limit the usefulness of this detector type for long-lived satellite missions such as XMM.

A second form of continuous UV readout (to date, perhaps the most highly developed) is the photoionisation chamber (fig. 2.30b), first proposed by Policarpo (1978) and actively pursued by X-ray astronomers at Columbia (Hailey *et al.*, 1983) and particle physicists at CERN (Charpak, 1982) and elsewhere.

In the Columbia detector, UV light leaving the noble gas cell of the GSPC enters a second gas cell, a multi-wire proportional counter containing a UV-ionisable organic vapour mixed with a conventional counting gas such as argon–20% methane (P20) or argon–20% isobutane. Provided the emission spectrum of the GSPC's noble gas is well-matched to the absorption spectrum of the MWPC additive, the UV light flash in the former is efficiently transformed into an electron cloud in the latter. The original work of Policarpo (1978) used triethylamine (TEA) gas to convert the scintillation light from krypton. Later researchers (Anderson, 1981; Hailey *et al.*, 1983), using xenon for reasons of X-ray stopping power (section 2.2.3), have found tetrakis (diethylamino) ethylene (TMAE – colloquially 'Tammy') with an ionisation potential of 5.36 eV, to be a suitable additive gas. Typical concentrations of TMAE are ~ 0.5 mbar in a total counting gas pressure of ~ 1 bar. Energy information is abstracted from the MWPC anode plane; position information, from centroid computation in orthogonal cathode strips (section 2.4.1.2).

Large area (up to 5 cm active diameter) photoionisation GSPCs have been built (Hailey *et al.*, 1983), with energy resolutions approaching the Fano factor limit ($\Delta E/E = 8\%$ at 6 keV) and spatial resolutions ($\Delta x = 2.2/E^{\frac{1}{2}}$ mm fwhm) comparable with the best GSPC/MCP results so far reported. The photoionisation GSPC, moreover, is the only imaging GSPC to have been tested in space. Prototypes were flown on sounding rockets in 1981 and 1983 to observe the Cygnus Loop (Ku *et al.*, 1983) and SN1006 (Vartanian *et al.*, 1985) supernova remnants. Results from the latter flight are shown in figs. 2.31a and b. Although statistically limited because of the short sounding rocket observing time, this map and this spectrum currently represent the 'state of the art' in imaging X-ray spectrophotometry. For long-term satellite operation, the susceptibility of a large organic molecule like TMAE to in-orbit radiation damage and cracking (section 2.3.3) remains to be quantified. Since the equilibrium vapour pressure of the additive in the MWPC gas mixture, and hence its UV absorption coefficient, are strong functions of temperature, active control of the chamber temperature would be required in the satellite environment.

The use of proportional counters containing UV sensitive vapours and liquids in fields other than astronomy (e.g. Cherenkov ring imaging) is reviewed by Anderson (1985). Low-pressure ($\sim$5 Torr) cells filled with TMAE and isobutane have been suggested as large-area UV detectors for astronomy (Michau *et al.*, 1986). The same authors found that no signal could be abstracted from condensed or adsorbed TMAE layers.

The most recent development in GSPC imaging uses a readout element which may, perhaps, be best regarded as a descendant of the Anger Camera. The production (by Hamamatsu Photonics KK in Japan; Kume *et al.*, 1985, 1986) of multi-anode photomultiplier tubes effectively packages an array of PMTs within a single, compact envelope. In one early design, a number of conventional 'Venetian blind' mesh dynodes was followed by a four-quadrant anode (fig. 2.22a); in another, the last dynode in the multiplier chain was a resistive sheet. In both these designs, the avalanche location was established by an appropriate charge division algorithm. Anger Camera algorithms (fig. 2.29a) are used in the more recent, more highly divided (3×3 anode array), PMTs.

Application of the multi-anode PMT to X-ray astronomy has been extremely rapid. Smith *et al.* (1987) describe a 0.1–10 keV low-energy GSPC (LEGSPC) for the focus of a flux concentrator (section 1.4) to be carried on the Italian national X-ray astronomy satellite SAX in the mid-1990s (SAX will also carry a high-pressure (5 bar Xe) Anger Camera GSPC for coded aperture imaging in the 4–100 keV band; Manzo *et al.*, 1985). Spatial resolutions measured with a nine-anode PMT in the LEGSPC development phase follow the promising trend

$$\Delta x = 2.2/E^{\frac{1}{2}} \text{ mm fwhm}$$

albeit over rather small (5 mm diameter) fields (Smith *et al.*, 1987). Industrial development of multi-anode PMTs with active areas much larger than the currently available 4×4 cm^2 tubes will be required for missions like XMM. The SAX LEGSPC will be a 'driftless' detector: that is, the drift region of the conventional imaging GSPC will be eliminated, X-rays then being directly absorbed in a scintillation region. Several small sources of energy blur (loss of primary electrons to the X-ray window, to gas

Fig. 2.31. (a) 0.1–1.4 keV map of the supernova remnant SN1006, obtained on a sounding rocket flight of the Columbia photoionisation chamber GSPC (Vartanian *et al.*, 1985). The image contains $\sim$ 700 counts. (b) X-ray spectrum of SN 1006 (Vartanian *et al.*, 1985). The crosses are measured fluxes. The histogram is a best-fit power law plus thermal plasma model (Appendix B). The excess flux at 0.59 keV is identified with O VII and O VIII line emission. (*Courtesy K. S. K. Lum.*)

ⓐ

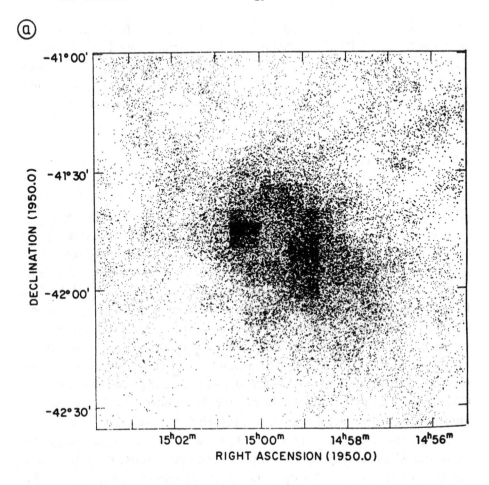

ⓑ

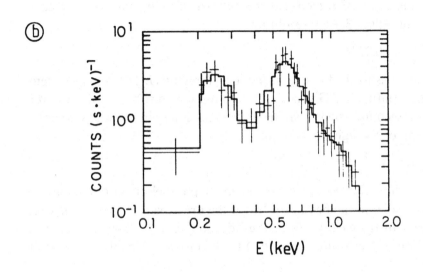

impurities and to the mesh separating low-field and high-field regions) arise in the drift region. Position resolution is also degraded by diffusion effects therein (eq. 2.16: Simons *et al.*, 1985a). Elimination of the drift region does make the gain of the GSPC dependent on the depth of X-ray penetration into the gas. This effect, however, can in principle be compensated for by measurement of the depth dependent UV burstlength.

2.5.3   *Electron counting*

The technique of electron counting was first described by Siegmund *et al.* (1982) at MSSL. In a conventional GSPC, the integrated light yield of all N accelerated photoelectrons is taken as an index of the incident X-ray energy. In an electron counting detector, N is determined directly by counting the number of individual pulses in the train of emitted light. In a conventional GSPC, the scintillation medium is usually xenon, an expensive gas which is susceptible to contamination and emits in the UV. The filling gas used in electron counting detectors is basically argon–($\sim 5\%$)methane, with additions of carbon dioxide ($\sim 5\%$ – see below) and nitrogen ($\sim 7\%$ – to provide increased light output). Light is emitted mainly from the excited states of atomic argon (7000–8000 Å) and the second positive band of nitrogen (3000–4000 Å). As in a conventional GSPC, the energy resolution attainable by electron counting should approach the Fano factor limit (eq. 2.10) when N is large.

Energy resolution measurements were made using a parallel-plate avalanche chamber (fig. 2.1d) with a conductive, transparent tin oxide anode deposited on a fibre optic faceplate. A fast photomultiplier tube viewed the emitted light. Fast discriminator and integrator electronics (Siegmund *et al.*, 1983b) recorded the number of individual peaks in each PMT output pulse. Energy resolutions

$$\Delta E/E = 0.22/E^{\frac{1}{2}}$$

were obtained using field strengths in the drift region (E_d) of only a few tens of volts per centimetre. Given the finite pulse pair resolution of the PMT ($\sim$ few nanoseconds), such weak fields were necessary in order to maximise the duration of the light pulse train

$$\tau = \sigma_z/W$$

Here, σ_z is the fwhm extent of the primary photoelectron cloud, in the direction parallel to the field, as it leaves the drift space (eq. 2.5) and W is the electron drift velocity. With the addition of the 'slow' gas carbon dioxide (Fraser and Mathieson, 1986) to the mixture, τ could be made of order 1 μs, and pulse pile-up avoided for X-ray energies up to 2 keV ($N < 80$).

For imaging studies, the photomultiplier tube was replaced by a microchannel plate based image intensifier with an S20 optical photocathode (Siegmund *et al.*, 1983a). Spatial resolution

$$\Delta x = 0.4 \text{ mm fwhm}$$

over a 2.5 cm diameter field was obtained by 'centroiding' the individual light flashes, despite a rather poor intensifier quantum efficiency. These results, however, were obtained at much higher drift field strengths ($E_d = 300$ V/cm) in order to minimise lateral diffusion effects (eq. 2.4).

Good energy resolution and good spatial resolution have never, in fact, been achieved simultaneously in an electron-counting detector. There are both fundamental (the requirement for the ratio of longitudinal and transverse diffusion coefficients to be large, when, for most gases at most field strengths, $D_L/D < 1$ (Fraser and Mathieson, 1986) and technical (the need for ultra-fast PMT response) reasons why this goal may remain extremely difficult to achieve over useful X-ray bandwidths. For the time being, at least, X-ray astronomers have moved on, to consider the properties of multistep avalanche chambers (next section). Active research into imaging the light from proportional counter avalanches continues in other fields (Gilmore *et al.*, 1983; Charpak, 1985).

2.5.4 *Multistep avalanche chambers*

In all the conventional (avalanche) proportional counters described so far in this chapter, there has been an implicit conflict between the requirements of imaging and spectroscopy. Good spatial resolution demands high gas gain G in order to 'drive' the position readout (eq. 2.15), but because the avalanche variance f (eq. 2.9) is, generally, an increasing function of G, optimum energy resolution is obtained at low gains. The prediction of Alkhazov (1970), extended by Sipila (1979), that f can be made close to zero in an optimised (parallel-plate) electric field, would therefore appear to be of only academic interest, since the field strengths required (~ 5 V/cm Torr in Ne–Ar) would lead to impracticably low (~ 50) gas gains in any real detector.

Multistep chambers were first developed for high count rate applications in particle physics (Breskin *et al.*, 1979; Breskin and Chechik, 1984). The multistep geometry featured two high-field regions; a preamplification or first avalanche region (between grids G1, G2 in fig. 2.1f) and a second avalanche region (between grid G3 and the readout anode), separated by a low-field transfer region. With the two avalanche regions operated in series, and with the transfer region alleviating the problem of multiple pulsing due to photon feedback, very high overall gains could be stably obtained.

The attraction of the multistep geometry to X-ray astronomers should be immediately obvious. One may abstract from the preamplification region (grids G1 and/or G2), operating at low gain, an energy signal which is little affected by avalanche fluctuations, while the high gas gain necessary for imaging ($> 10^5$) can be produced independently, in the second avalanche region.

The Penning Gas Imager (PGI: Schwarz and Mason, 1984) was the first multistep detector built with the goal of combining good energy resolution and good spatial resolution. The name comes from the filling gas used, argon–0.5%C_2H_2, one of the Penning gases investigated by Sipila (1976). Since the acetylene admixture has an ionisation potential lower than the energy of the photons emitted by excited argon atoms, it may be photo- or collisionally ionised by the argon. The process:

$$Ar^* + C_2H_2 \rightarrow Ar + (C_2H_2)^* + e^-$$

is an example of the Penning (1934) effect. The extra ionisation increases the first Townsend coefficient α (section 2.2.2) and decreases both w, the average energy needed to create an ion pair, and F, the Fano factor (section 2.2.1) of the gas. While these properties make Penning counting gases of interest in their own right (Sephton et al., 1984), the ability of the argon component to photoionise the acetylene admixture was the critical factor which led to the adoption of Ar–C_2H_2 by the MSSL researchers.

In any multistep chamber, there is the problem of how to drift electrons from the high-field preamplification region through an intervening wire mesh to a region of much lower field strength. In a conventional gas mixture, the transfer efficiency goes as the (small) ratio of field strengths in the two regions. According to Schwarz et al. (1984), very much enhanced electron transfer efficiencies (up to 20%) result from the use of an appropriate Penning gas, because of the ionising effects of the argon UV light transmitted through the grid. These authors in fact claim that the use of a Penning gas is essential for successful operation of multistep counters (the results of Breskin et al., 1979, were obtained in an argon–acetone mixture).

Although in principle the energy resolution of the PGI should approach that of a perfect GSPC (eq. 2.10), the best results so far reported, using an Ar–2%C_2H_2 gas mixture, follow the curve:

Fig. 2.32. Exploded view of the Penning Gas Imager (Schwarz and Mason, 1985). (*Courtesy H. Schwarz. © 1985 IEEE.*)

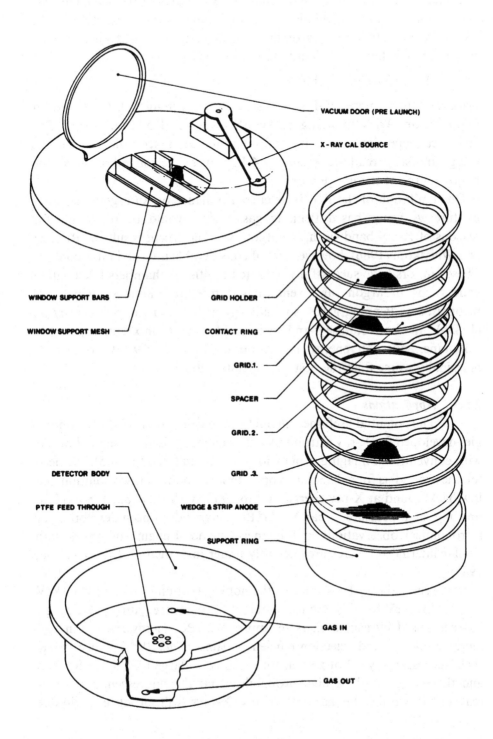

$$\Delta E/E = 0.29/E^{\frac{1}{2}}$$

intermediate to the best MWPCs and GSPCs. Further improvement may be possible if gain modulation at field boundaries can be reduced by changing the detailed shape of the field-defining grids. Excellent spatial resolution has, however, already been obtained:

$$\Delta x = 0.34/E \, \text{mm fwhm}$$

using a 12 cm diameter PGI (shown schematically in fig. 2.32) intended for flight on an Aries sounding rocket (Schwarz and Mason, 1985). The reciprocal dependence on E indicates dominance of electronic noise in the wedge-and-strip readout (section 2.4.1.2) over electron diffusion as the major source of the image blur.

One limitation of the PGI is its current reliance on an argon-based gas mixture. A xenon-based Penning gas, which would be preferable for reasons of X-ray bandwidth, is not known. The recent results of Ramsey and Weisskopf (1986), however, are of considerable interest in this context. These authors have successfully operated multistep chambers filled with a wide range of argon- and xenon-based mixtures, none of which are recognised as Penning gases, so challenging the claims of Schwarz *et al.* (1985b). In fact, Ramsey and Weisskopf's 'best' mixture, argon–5% isobutane, gives an energy resolution of 13% at 5.9 keV, only very marginally worse than the best PGI results quoted above.

2.5.5   *Applications*

The potential usefulness in nuclear medicine of the GSPC technologies developed for X-ray astronomy has been recognised for some time. As with conventional imaging proportional counters (section 2.4.5), the need 'on the ground' is for a detector to operate with good energy resolution and low background at X-ray energies up to 100 keV (Sims *et al.*, 1984). Non-imaging, high-pressure (5 bar Xe) GSPCs have indeed been developed for astronomical observations in this energy band. Imaging detectors with similar internal pressures are currently under investigation (Manzo *et al.*, 1985).

At energies above the K edge of the noble gas counter filling (34.65 keV for Xe; 14.3 keV for Kr), the technique of fluorescence gating ('K gating') has been used by astronomers to improve GSPC energy resolution and increase background rejection efficiency (Sims *et al.*, 1983). Since the K shell fluorescence yield of xenon, for example, is large (0.875: see fig. 2.2) and there is generally a high probability that the fluorescent photon is reabsorbed within the gas cell, many X-rays will produce a double signature in a xenon-filled GSPC which will distinguish them from background events (K gating can, of course, be used just as well in a conventional xenon-filled proportional counter; Ramsey *et al.*, 1985). The energy of the fluorescent event is, moreover, known precisely, so that the energy of the original X-ray can be inferred with a resolution rather better than that of an equivalent, ungated counter.

High-pressure (~ 7 bar) versions of the PGI have recently been proposed for muscle diffraction pattern imaging in biology (Schwarz *et al.*, 1985b). Bateman *et al.* (1984) have described the application of multistep techniques to radiography.

3

Microchannel plates

3.1 Introduction

Microchannel plates (MCPs) are compact electron multipliers of high gain and military descent which, in their two decades as 'declassified' technology (Ruggieri, 1972), have been used in a wider range of particle and photon detection problems than perhaps any other detector type.

A typical MCP consists of $\sim 10^7$ close-packed channels of common diameter D, formed by the drawing, etching and firing in hydrogen of a lead glass matrix. At present, the most common values of D are 10 or 12.5 µm, although pore sizes as small as 2 µm have begun appearing in some manufacturer's literature. Each of the channels can be made to act as an independent, continuous-dynode photomultiplier. Microchannel plates (or channel multiplier arrays or multichannel plates, as they are sometimes known) are therefore used, in X-ray astronomy as in many other fields, for distortionless imaging with very high spatial resolution.

The idea of replacing the discrete dynodes (gain stages) of a conventional photomultiplier (Knoll, 1979) with a continuous resistive surface dates from 1930 (Ruggieri, 1972). It was only in the early 1960s, however, that the first channel electron multipliers (CEMs), consisting of 0.1–1 mm diameter glass or ceramic tubes internally coated with semiconducting metallic oxide layers, were constructed in the USSR (Oshchepkov et al., 1960) and United States (Goodrich and Wiley, 1962). Somewhat later, parallel-plate electron multipliers (PPEMs) were developed with rectangular apertures more suited to the exit slits of certain types of spectrometer (Spindt and Shoulders, 1965; Nilsson et al., 1970). The principles of continuous-dynode multiplier operation are illustrated in fig. 3.1. Fig. 3.2 shows the range of modern CEMs offered by one manufacturer; all have curved channel geometries in order to suppress noise due to ion feedback, discussed in the context of MCPs in section 3.3.2. Such 'channeltrons' have been very widely used in space astronomy, from the mid-1960s onwards

(Smith and Pounds, 1968; Timothy and Timothy, 1969), as rugged, windowless detectors of both charged particles and photons. More recently, CEMs with coned apertures (fig. 3.2) have been used in conjunction with paraboloidal X-ray and XUV concentrators (section 1.4) on the Copernicus (Bowles *et al.*, 1974) and Apollo–Soyuz (Margon and Bowyer, 1975; Lampton *et al.*, 1976a) missions.

The first imaging multipliers were produced by the straightforward but laborious technique of stacking together individual straight channel CEMs (Wiley and Hendee, 1962) or 'spiraltrons' (curved channel multipliers with a right cylindrical geometry: Somer and Graves, 1969), into large arrays. Eventually, however, investment directed towards the production of low light level ('night vision': Schagen, 1971) image intensifiers for the military led to the development, by several manufacturers, of the 'two draw, etchable core' microchannel plate process illustrated in fig. 3.3 (Washington *et al.*, 1971; Asam, 1978). The concept of the spiraltron array has recently been revived with the development of MCPs with 'off-axis' channels, in which the individual fibres in the MCP boule each consist of a number of helically twisted channels (Timothy *et al.*, 1986).

Night vision image tubes continue to provide the major market for most

Fig. 3.1. Operation of a continuous-dynode electron multiplier. X-rays (or other photons or particles) incident at the low potential end of the multiplier release photoelectrons from its semiconducting surface. These photoelectrons are accelerated in vacuum by the applied electric field ($V_0 \sim 1.5$–$4\,$kV), strike the opposite wall of the multiplier and, provided the secondary electron emission coefficient of the surface is greater than one, release secondaries and initiate a cascade. G ($> 10^8$) is the multiplier gain in electron/photon. The same operational principles apply to both parallel-plate electron multipliers (PPEMs: orthogonal section at right), in which the semiconducting layer might be amorphous silicon (Nilsson *et al.*, 1970) and tubular channel electron multipliers (CEMs: left), usually constructed from reduced lead oxide glass. The average linear gain in both geometries depends, not on the diameter D, but on the dimensionless ratio L/D (Adams and Manley, 1966). This is the physical principle which made possible the development of MCPs by the miniaturisation of large bore CEMs.

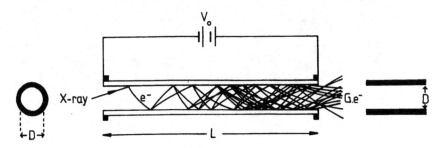

of the MCP manufacturers listed in Table 3.1, although, for some of these companies, the mass production of large-area ($>100 \, \text{cm}^2$) channel plates for incorporation in fast cathode ray tubes (Wiza, 1979) has recently assumed increasing importance. By comparison with either of these markets, the demand from X-ray astronomers (and all other scientific channel plate users) is minute.

Minute, but not unimportant. Many developments in channel plate design, particularly those related to channel geometry (section 3.2), have in

Fig. 3.2. Modern CEMs. *Scale*: the tube diameter of the smallest multiplier shown is about 1 mm. (*Courtesy Philips Components (formerly Mullard) Ltd.*)

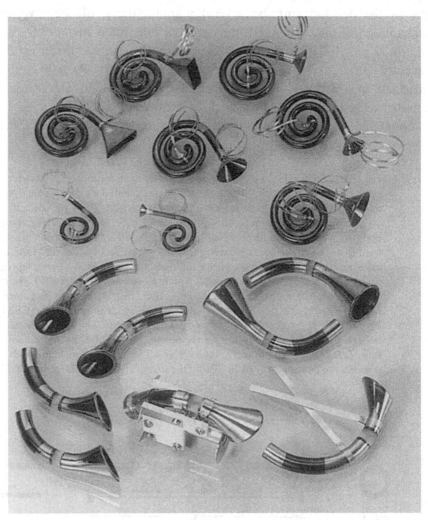

fact been in response to the needs of astronomers working in the X-ray, XUV and ultraviolet bands. Just as the main impetus in the field of gas detectors (Chapter 2) has been provided by the particle physicists, so advances in microchannel plate technology have owed a great deal to the desire of the space astronomy community for large-area, high-gain, low-noise photon detectors of high spatial resolution.

Fig. 3.3. Steps in the manufacture of microchannel plates (Mullard technical information, 1981). A hollow billet of lead oxide cladding glass is mechanically supported by the insertion of a rod of etchable core glass and then pulled through a vertical oven, producing a 'first draw' fibre of diameter ~1 mm. Lengths of first draw fibre are then stacked (usually by hand) in a regular hexagonal array which is itself drawn to produce a hexagonal 'multifibre'. Lengths of multifibre are stacked in a boule and fused under vacuum. The boule is sliced and the slices are polished to the required MCP thickness and shape. The solid core glass is then etched away, leaving the channel matrix to be fired in a hydrogen oven to produce a semiconducting surface layer with the desired resistance and secondary electron yield. (*Courtesy Philips Components (formerly Mullard) Ltd.*)

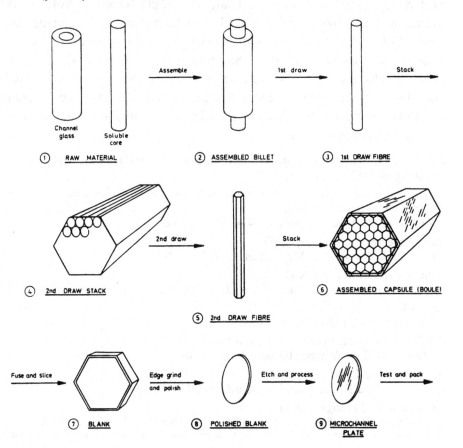

In the next section, we shall describe the physical composition of microchannel plates, together with the various geometric formats now available. Section 3.3 outlines the principles of X-ray detection with MCPs, while section 3.4 concentrates on imaging, with reference to a number of flight instruments for X-ray and XUV astronomy. Finally, section 3.5 considers the many fields in which astronomical research into channel plates could be, or already has been, usefully applied.

3.2   MCP composition and format

Microchannel plates are manufactured (fig. 3.3) from lead glasses whose densities lie in the range 3.3–4.3 g/cm^3. The bulk composition of the glass varies from manufacturer to manufacturer, and is under continual development by each one of them, but its principal constituents are always PbO and SiO$_2$, with alkali metal oxides making up several per cent by weight (Mackenzie, 1977; Wiza, 1979: see section 3.3.6). It is this bulk composition (up to 48% lead by weight) which is responsible for the MCP's hard X-ray, γ-ray and energetic charged particle response. Soft X-rays interact with a ~ 100 Å thick surface layer which is silica-like, having been depleted of lead and enriched with potassium and carbon during the final stages of processing (Hill, 1976; Siddiqui, 1977; Fraser, 1982). It is the secondary electron yield of this activated layer which determines the MCP gain (Trap, 1971). Compositional changes within this layer are generally held to be responsible for gain lifetime effects (section 3.3.2.2) and changes in plate resistance (below).

Table 3.1. *Microchannel plate manufacturers (M) and commercial suppliers of complete MCP detectors (D)*

Philips Components Ltd. (M)
4 New Road, Mitcham, Surrey CR4 4XY, England

Galileo Electro-Optics Corp. (M)
Galileo Park, Sturbridge, Massachusetts 01518, USA

Hamamatsu Photonics KK (M, D)
314–15 Shimokanzo, Toyooka-Village, Iwata Gun, Shizuoka-Ken, Japan

Varian LSE Division (M)
611 Hansen Way, Palo Alto, California, USA

ITT Electro-Optical Products Division (M, D)
PO Box 3700, Fort Wayne, Indiana 46801, USA

Surface Science Laboratories (D)
4151 Middlefield Road, Palo Alto, California 94303, USA

Instrument Technology Ltd. (D)
29 Castleham Road, St. Leonards-on-Sea, E. Sussex TN38 9NS, England

As the final stage in the production process, the MCP is electrode coated on both faces so that a bias potential difference V_0 can be applied *in vacuo* (at pressures $< 10^{-5}$ Torr) along the length of the channels. Conventionally, the MCP input face is operated at a high negative potential and the charge collector at the output face is earthed. The electrode material (usually a nickel-based alloy such as inconel ($Ni_{10}Cr_2Fe$) or nichrome ($Ni_7Cr_2Fe_3$)) is vacuum evaporated at 45° to the plate normal, giving a one channel diameter penetration ('1D end spoiling') into the channels. This electrode penetration determines the soft X-ray detection efficiency at large angles of photon incidence, θ (section 3.3.3) and influences the lateral spreading of the charge cloud leaving a working channel (section 3.4).

Standard night vision MCPs are ~ 25 mm diameter discs, usually with a solid glass rim, whose straight channels of circular cross-section are hexagonally packed to a very high degree of uniformity (fig. 3.4). Such MCPs have an open area fraction, A_{open}, of around 63%. Square channel MCPs (Asam, 1978) have a larger open area than circular channel MCPs of the same minimum septal thickness. Night vision intensifiers incorporate single channel plates with channel length-to-diameter ratios L/D in the

Fig. 3.4. Scanning electron micrograph of part of an MCP with 12.5 micron diameter channels.

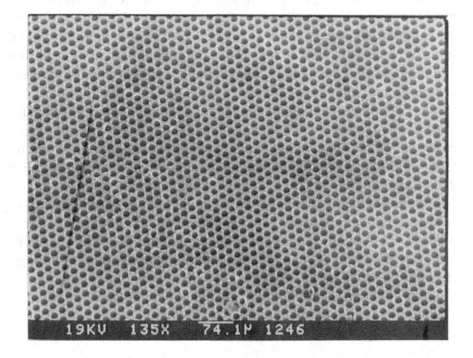

19KV   135X   74.1P 1246

range 40:1–60:1. Monte Carlo models of channel plates as current amplifiers (Guest, 1971) show that these are the L/D values which result in
minimum spatial variation of gain. They are not optimum for single
photon counting (section 3.3.2.1). The manufacturing process described in
fig. 3.3 is, however, very flexible: a variety of unconventional plate formats
are obtainable from at least some of the manufacturers listed in Table 3.1,
usually at the penalty of cost. These include MCPs with curved channels
(section 3.3.2.1), 'thick' MCPs with large L/D ratios ($80:1 < L/D < 175:1$)
and plates of large area. The largest MCPs of any kind so far manufactured
(for fast cathode ray tube and flat screen display applications) are rect-
angles $\sim 14 \times 18$ cm^2; the AXAF High Resolution Camera (HRC: section
3.4.2), incorporating 10×10 cm^2 plates, is the largest MCP detector
currently under development for X-ray astronomy. Rectangular MCPs
have been abutted to form strip detectors 36 cm long (Asam, 1978). Such
detectors may prove useful for transmission grating spectroscopy in a
number of future satellite missions. Central holes can be cut in circular
channel plates to form annular detectors (Weissenberger *et al.*, 1979; Reme
et al., 1987). Funneling of the channels, by adding an extra chemical etch to
the manufacturing sequence, leads to open area fractions, principally for
electron detection, in excess of 90% (Woodhead and Ward, 1977).

Since the focal surfaces of the grazing-incidence optics used in X-ray
astronomy are often highly curved, a sensitivity advantage accrues to any
telescope incorporating a detector of approximately matching curvature.
To gain this advantage, boule slices (fig. 3.3) can be spherically ground
(prior to the etching of the channels) to form plano-concave (input face
curved, output face flat) and bi-concave (both surfaces curved to the same
radius) microchannel plates (Fraser *et al.*, 1983; Timothy, 1985). Fig. 3.5
shows the bi-concave Mullard MCPs developed for the ROSAT Wide
Field Camera (section 3.4.3; Barstow *et al.*, 1985) which are curved to a
radius of 165 mm. Calculation shows that in this case a sensitivity advantage of ~ 2.5 times results from the use of a curved detector, rather than a
flat detector in focus only at the centre of the field-of-view.

Irrespective of format, channel plates normally have *in vacuo* resistances
of several-hundred megohms. The resistance of an individual channel is of
order $10^{14} \Omega$. Resistances within a batch of geometrically identical MCPs
may vary by up to $\pm 50\%$ (fig. 3.6), reflecting the influence of single
channels of anomalously low resistance when very many channels are
connected in parallel. Fig. 3.6 also illustrates the non-ohmic variation of
channel plate resistance with bias voltage V_0, originally noted by Rager and
Renaud (1974), which is simply a consequence of the heating of the plate by

the current passing through it (Pearson *et al.*, 1987). Channel plate glasses have temperature coefficients of resistance of about $-1.5\%/\text{K}$, a factor which should constrain the design of high-voltage power supplies for satellite-borne detectors, which may have to cope with significant temperature excursions.

The ten plates represented in fig. 3.6 were later vacuum baked for 48 hours at a temperature of 275°C. This processing, desirable for the long-term stability of MCP operation, particularly in the presence of a hygroscopic photocathode material, increased plate resistances by 15–30%. Siddiqui (1979) has observed 38–50% resistance increases after somewhat higher temperature bakes. These increases are caused by the desorption of gases (Morgan, 1971) from the channel walls and are irreversible after

Fig. 3.5. Microchannel plates. *Back right*: 53 mm diameter conventional MCP. *Front right*: 36 mm diameter conventional MCP. *Front left*: plano-concave MCP with 12.5 μm channels and variable thickness, such that $L/D = 120{:}1$ (centre) and 175:1 (edge). *Back left*: 55 mm diameter ROSAT Wide Field Camera MCP. Patterns on the plate surfaces are reflections of an overhead mesh, introduced to highlight curvature.

bakeout above a transition temperature in the range 215–250°C (Rager *et al.*, 1974; Siddiqui, 1979).

The product of an MCP's resistance and the capacitance between its input and output faces determines its gains at high output count rates (section 3.3.2.3); since the relative permittivities of MCP lead glasses range from 7.8 to 9.9 (Mackenzie, 1977), the latter parameter usually lies between 20 and 150 pF.

Fig. 3.6. *In vacuo* resistance versus bias voltage V_0, for a batch of ten (unbaked) 36 mm diameter MCPs. Channel diameter $D = 12.5$ μm, length-to-diameter ratio $L/D = 40:1$. Resistances nominally obey a L/(plate area) scaling law, so that 120:1 MCPs otherwise identical to these have resistances in the range 500–700 MΩ.

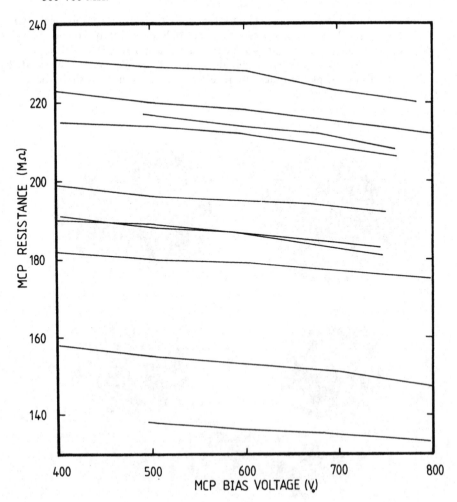

3.3 Physical principles of microchannel plate operation

The following sections describe the fundamental physical properties of microchannel plate X-ray detectors: gain, quantum detection efficiency, dark noise and the like. Because microchannel plates only operate stably at pressures below $\sim 10^{-5}$ Torr, measurement of these properties can only properly be made in specialised test chambers constructed to ultra-high-vacuum standards. For most of the other X-ray detectors described in this monograph a test vacuum is required in the first instance to reduce attenuation of the calibration source.

Figs. 3.7a–c describe one of the microchannel plate test chambers currently used by the X-ray astronomy group at Leicester. Similar facilities have been described by Bjorkholm *et al.* (1977) and, for XUV detector calibration, by Malina *et al.* (1985).

3.3.1 *X-ray interactions*

At the energies of principal interest here, X-rays interact with the channel plate glass and electrodes, and with any associated photocathode material (section 3.3.3), via the photoelectric effect. For X-ray energies below about 5 keV (Fraser, 1982) detection proceeds in a 'single channel' mode. No significant fraction of the X-ray beam entering a given channel penetrates the channel wall to illuminate the next channel in the structure. The range of the most energetic photoelectrons is small compared with the minimum septal thickness

$$s_{min} = p - D$$

where p is the inter-channel pitch, equal to 15 μm in standard MCPs with 12.5 μm diameter channels. At higher energies the effect of 'multiple channel crossing' of the incident beam becomes significant (section 3.3.3).

The escape of the primary photoelectron and/or Auger electron from the channel wall may be accompanied by the ejection of one or more secondaries. The energy required to excite one internal secondary electron, w (eq. 2.1), is 10 eV for channel plate lead glass (Fraser, 1983a), which thus compares favourably with gaseous detection media (section 2.2.1) in terms of the number of photoelectrons produced per kilovolt X-ray energy. The escape length for secondary electrons in lead glass, however, is only 33 Å (Fraser, 1983a), so that the most probable number of electrons escaping into the channel is always one, independent of X-ray energy (Fraser and Pearson, 1984). 'Bare' MCPs, in which the glass and electrode surfaces prepared by the manufacturer act as photocathodes, thus possess very limited energy resolution (section 3.3.4).

3.3.2   MCP gain
3.3.2.1   *Linear and saturated operation*

For single photon or charged particle counting, MCPs should be used in one of a number of high gain configurations which produce a 'saturated' (i.e. peaked) output pulse height distribution. These configurations are illustrated in fig. 3.8.

A large peak (modal) gain G (of order 10^6–10^8), accompanied by a small pulse height fwhm $\Delta G/G$ ($< 50\%$), permits a measure of discrimination between X-ray signals and internal noise (section 3.3.6), while maximising the electronic signal-to-noise ratio and minimising the dynamic range to be dealt with by position encoding electronics acting on the output pulse (section 3.4).

In the low voltage, low gain linear or d.c. régime ($G < 10^4$), the output pulse height distribution of a single MCP is quasi-exponential in nature (Adams and Manley, 1966; Guest, 1971; Csorba, 1980; Eberhardt, 1981). A cascade electron may follow a great variety of trajectories between collisions with the channel wall. The electron energy, and hence the secondary electron yield at each collision, is ill-defined and a large spread of gains results. First generation night vision MCPs ($L/D = 40:1$; $D = 40\,\mu$m) could not be made to display the output charge saturation exhibited by CEMs before increasing V_0 caused the onset of ion feedback. Ions produced by electron collisions with residual gas molecules travelled back along the active channel and initiated after-pulses of large magnitude. Two first generation plates, however, could be operated in series (fig. 3.8b) to produce feedback-free, saturated gains in excess of 10^7. This is the so-called 'chevron' configuration, first patented by Bendix Research Laboratories

Fig. 3.7. University of Leicester X-ray astronomy group MCP detector test facility. (a) Detector body. The compression block, spacer and internal detector body are made from Macor machineable ceramic. Leaf spring: Cu–Be sheet. Graded density (GD) cathode frames (Smith *et al.*, 1982): G10 fibreglass. All other components: stainless steel. Electrical contact to the MCPs is via three gold-plated copper electrodes. The assembly is held together by four screws through the top plate. (b) Detector carriage, providing ± 25 mm linear translation and $\pm 60°$ rotation of the detector assembly relative to a fixed X-ray beam. (c) Modular, oil free stainless steel test chamber. Main chamber evacuated by turbomolecular pump (TMP); X-ray source chamber, by ion pump/sorption pump combination. A single-wire proportional counter reference detector can be lowered into the collimated X-ray beam from a filtered (Bjorkholm *et al.*, 1977) coated anode source. The distance from the filter wheel to the MCP detector is 1.5 m. The conventional X-ray source is used for 0.1–3 keV measurements. It can be replaced by radioisotope emitters or UV lamps for measurements at higher or lower photon energies. (*Courtesy J. F. Pearson.*)

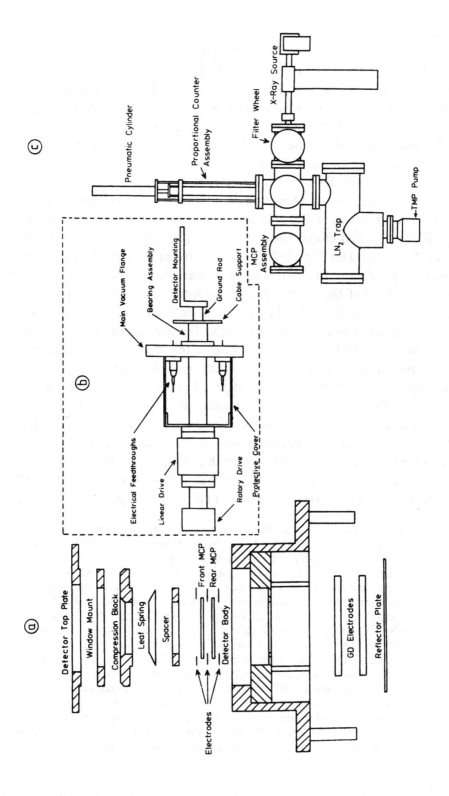

(later Galileo Electro-Optics) in 1968 (Ruggieri, 1972). Many reviewers cite the paper by the Bendix researchers Colson *et al.* (1973) as the first report of a two-stage multiplier in the scientific literature, although that distinction probably belongs to Parkes and Gott (1971), reporting results obtained at Leicester. Other early accounts of chevron detectors are given by Catchpole and Johnson (1972) and Washington (1973).

Experiments with single straight channel MCPs of increased L/D ($>75{:}1$) also led to peaked pulse height distributions (Chalmeton and Eschard, 1972) but did not completely eliminate ion feedback. It has since been shown that a pulsed bias voltage suppresses ion feedback in a single 160:1 straight channel MCP (Gatti *et al.*, 1983). Attempts to produce MCPs with angled electric fields within straight channels met with only limited success (Timothy, 1974). Curved channel MCPs with large L/D ratios did produce feedback-free gains of order 10^6, one order of magnitude less than those of the typical chevron (Boutot *et al.*, 1976; Timothy and Bybee, 1977; Timothy, 1981, 1985). The 'C' or 'J' channel curvature in such plates suppresses ion feedback in similar fashion to the spiral or helical curvature of a CEM, a tilted electric field or the inter-plate gap and change in channel direction employed in chevron or Z-plate detectors (figs. 3.8b, a). That is, ion trajectories are limited geometrically, so that ion energies on collision with the channel wall are insufficient to initiate after-pulses.

Claims continue to be made to the effect that curved channel MCPs have superior ion feedback characteristics to all combinations of straight channel plates (Timothy, 1985). Other than examining MCP pulse shapes with very high time resolution (Hocker *et al.*, 1979; Oba and Rehak, 1981), the simplest way to detect ion feedback in any detector geometry is to look

Fig. 3.8. MCP detector geometries. *Symbols*: UVIS, ultraviolet/ion shield; M, field defining mesh; MCPi ($i = 1, 2, 3$), microchannel plate; CCMCP, curved channel microchannel plate; PSR, position-sensitive readout. (a) Cross-section of three stage 'Z' plate multiplier. θ_B is the bias angle of the channels. Plates shown separated by gaps of thickness d_G. Z stacks have been constructed with constituent plates mechanically pressed in contact or permanently bonded together (so-called laminated (Henkel *et al.*, 1978) or sandwich (Pearson *et al.*, 1988) configurations). For a Z plate with all three stages in contact, the most probable numbers of activated channels are three and seven in the second and third plates, respectively (Eberhardt, 1981). (b) Two stage chevron (or herringbone or tandem) configuration, with zero degree bias front plate. V_M, V_F, V_G, V_R, V_A are the potential differences between the various detector elements. Stacks of up to five straight channel MCPs (effectively a chevron plus a Z plate) have been reported (Firmani *et al.*, 1982; Crocker *et al.*, 1986). (c) Curved channel or 'C' plate. Annular focus electrodes have been used as an alternative to the field defining mesh in C-plate detectors (Timothy, 1985).

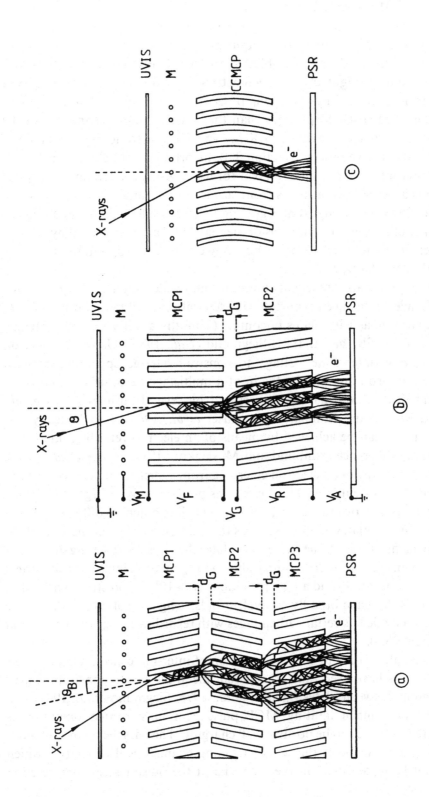

for a high-charge 'tail' in the output pulse height distribution. Using this criterion, the weight of the published evidence tends, if anything, to favour the various straight channel geometries. Certainly, MCPs with curved channels, produced by a shearing process (Timothy *et al.*, 1986), are more difficult and costly to produce than those with straight channels. All the MCP detectors so far flown in X-ray and XUV astronomy, and many of those under development (section 3.4), incorporate straight channels.

The front MCP of a chevron pair for use in X-ray astronomy usually has channels perpendicular to its front surface (a bias angle $\theta_B = 0°$: fig. 3.8b), since X-rays emerging from a grazing-incidence telescope do so along the surface of a cone. The channels of the rear MCP are then biased by 13–15°. Whether or not such relative bias angles are truly optimum is unclear (Parkes and Gott, 1971).

Inter-plate gaps are usually about 0.1 mm wide, but acceptable detector performance has been reported with separations of 1 mm (Firmani *et al.*, 1982) as well as with plates in contact (Timothy and Bybee, 1975; Henkel *et al.*, 1978; Eberhardt, 1981; Siegmund *et al.*, 1985). Use of an electron accelerating inter-plate voltage to constrain charge spreading between plates, and so minimise the fwhm of a multistage detector's output pulse height distribution, has been discussed by several authors (Wiza *et al.*, 1977; Rogers and Malina, 1982; Fraser *et al.*, 1983). Paradoxically, the same result can be achieved by means of an electron retarding inter-plate potential difference (Ainbund and Maslenkov, 1983; Fraser *et al.*, 1985). Since the energy of an electron emerging from a microchannel is coupled to the angle of its emission ('fast' electrons preferentially emerge close to the channel axis; Bronshtein *et al.*, 1979), a retarding potential difference of 25–50 V between plates actually reduces the illuminated area in the second (or subsequent) MCP. A chevron (or Z-plate) thus has two useful gain modes: a low gain mode ($G \sim 10^6$, comparable to the gain of a single curved channel MCP, V_G negative) and a high gain mode ($G > 10^7$, V_G positive in fig. 3.8b). These modes have broadly equivalent pulse height resolutions $\Delta G/G$. The low gain mode has certain advantages with regard to energy discrimination (section 3.3.4).

As an alternative to the manipulation of inter-plate electric fields, some researchers have used thin, transparent metal meshes to constrain charge spreading between plates (Henkel *et al.*, 1978; Siegmund *et al.*, 1985).

The mechanism limiting MCP pulse gain in saturated operation is generally believed to be positive wall charging. The passage of an avalanche progressively depletes the semiconducting channel wall of charge, which cannot be replenished on the timescale of the pulse transit time (section

3.3.7). Electron collision energies decrease as a retarding 'wall charge' electric field becomes established, until, near the channel exit, they are sufficient to eject secondaries from the glass only on a one-for-one basis. The charge cloud's magnitude stabilises or 'saturates'.

Zeroth-order calculations (Loty, 1971; Leonov *et al.*, 1980) based on the wall charge model predict a saturated gain dependence for a single MCP

$$G \propto V_0 D(L/D)^{-1}$$

This dependence is confirmed by a detailed numerical model (Fraser *et al.*, 1983). Numerically calculated gains (fig. 3.9) are in reasonable agreement with measurements made on both straight and curved channel single MCPs.

The following semi-empirical equation relates the peak gain G_C of a two-stage chevron detector to the gains of its constituent front (F) and rear (R) plates (Fraser *et al.*, 1983):

$$G_C = G_F^{1-\alpha} G_R N_C^\alpha \tag{3.1}$$

where N_C is the number of channels in the rear MCP illuminated by the charge cloud emanating from the front plate. Here, α is a parameter related to the channel diameter; α has the value 0.6 for plates with 12.5 μm diameter channels (Fraser *et al.*, 1983; Smith and Allington-Smith, 1986) and the value 0.8 for 25 μm channels. An expression analogous to eq. (3.1) has been derived by Siegmund *et al.* (1985) to describe the operation of three-stage multipliers.

The theoretical and experimental evidence presented by Fraser *et al.* (1983), indicates that pulse height fwhm should decrease with channel diameter. That is, plates with smaller pores should produce more tightly peaked pulse height distributions. Indeed, the 'narrowest' pulse height distribution so far reported (fig. 3.10) was measured using non-standard plates with $D = 8$ μm, $L/D = 175:1$.

Irrespective of channel geometry, minimal fwhm is always obtained when (i) the individual stages of the multiplier are independently operated in 'hard' saturation, i.e. when the plate bias voltages independently exceed a level $(V_0)_s$, where (Fraser *et al.*, 1983):

$$(V_0)_s = \{8.94(L/D) + 450\} \text{ V}$$

and (ii) the inter-plate potential difference(s) is (are) well chosen. While Timothy (1981, 1985) has made definitive claims for the superiority, in terms of fwhm, of single curved channel MCPs over their straight channel competitors, the best pulse height distributions reported from Z-plates, chevrons and curved channel MCPs of the same channel diameter are in fact rather similar; fwhm values around 30% have been achieved with 12.5 μm channels in all three plate configurations.

Fig. 3.9. Calculated single MCP (peak) gain versus bias potential difference V_0 (Fraser *et al.*, 1983). Note the changes in slope of the various curves; these indicate a transition from linear to saturated operation as V_0 is increased. The (mean) linear gain depends only on L/D for a given value of V_0; saturated gains also scale with channel diameter D.

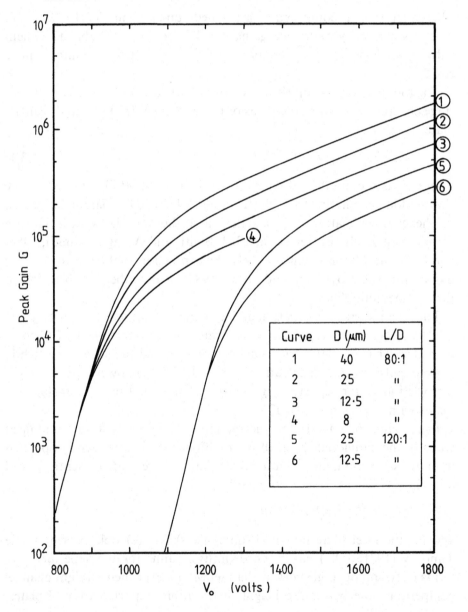

Curve	D (μm)	L/D
1	40	80:1
2	25	"
3	12·5	"
4	8	"
5	25	120:1
6	12·5	"

There are two situations when it is difficult to produce a peaked pulse height distribution with any channel plate detector. The first is when X-rays are incident at an angle θ to the channel axes less than θ_c, the critical angle for reflection from the channel glass (Fraser, 1982). Fig. 3.11 shows the broadening towards low charge levels of a C K (0.28 keV) X-ray pulse height distribution as θ is decreased. At small angles of X-ray incidence, electron avalanches can be initiated considerable distances from the channel entrance (up to $2\cot\theta$ channel diameters after one reflection) and so develop through only part of the potential drop across the plate. Such reflection effects are important when MCPs are used in conjunction with X-ray telescopes of long focal length. In the case of the AXAF High Resolution Camera (section 3.4.2), for example, angles of focused X-ray incidence will lie in the range 1.8–3.4°.

A second dispersion of event position, and hence of gain, occurs when the X-rays are sufficiently penetrative to release photoelectrons throughout the MCP bulk. Exponential pulse height distributions have been reported for 10–140 keV X-rays incident on a chevron (Bateman, 1977a) and for 1 MeV γ-rays in a saturable curved channel MCP (Timothy and Bybee, 1979). The mean X-ray absorption length in channel plate lead glass is

Fig. 3.10. X-ray induced pulse height distribution from a two-stage MCP detector with $D = 8 \ \mu m$, $L/D = 175{:}1$ channels (Fraser *et al.*, 1985).

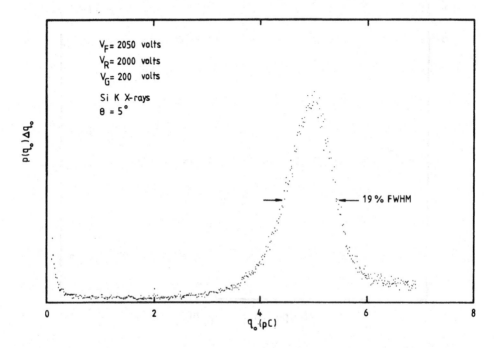

equal to the 1 mm thickness of a typical small pore MCP for energies around 50 keV (section 3.3.3). Wiza (1979) has reported the development of much thicker (10 mm) MCPs for γ-ray imaging which he claims to be

Fig. 3.11. Variation of chevron output pulse height distribution with C K X-ray angle of incidence, θ. When θ is less than $θ_c$ (6.6° for C K X-rays incident on MCP lead glass (Fraser, 1982), the effects of X-ray reflection broaden the distribution towards low charges. Front MCP: $D = 12.5$ microns, $L/D = 120{:}1$. A similar sequence is presented by Bjorkholm *et al.* (1977) for MCPs with $L/D = 80{:}1$.

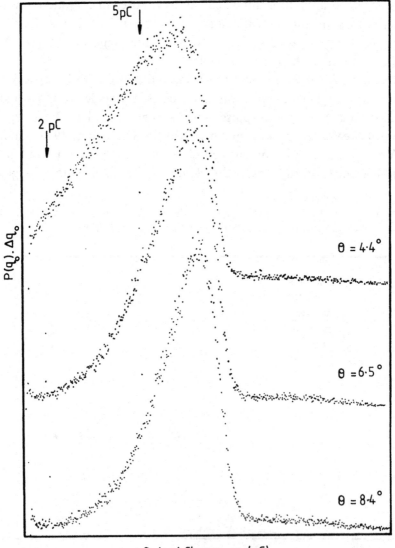

Output Charge q_o (pC)

saturable. Tests of these plates, which were byproducts of the EXOSAT MEDA and GSPC collimator development programme (sections 2.3.1, 2.5.1.2) at 511 keV have, however, resulted in pulse height distributions with the usual negative exponential form (McKee *et al.*, 1985).

3.3.2.2   *Gain fatigue*

The decrease of microchannel plate gain with abstracted charge is a widely reported but poorly understood phenomenon. The results of pre-1980 gain 'lifetests', conducted under a wide variety of experimental conditions, are summarised by Fraser (1984). More recent data is shown in figs. 3.12a and b.

Fig. 3.12a compares the results of the charge extraction lifetests conducted by Malina and Coburn (1984) and by Whiteley *et al.* (1984). The five Malina and Coburn data sets, all of which were obtained, at constant bias voltage, using chevron detectors of the same channel geometry ($D = 25\,\mu m$, $L/D = 40:1$), nevertheless exhibit a large scatter in relative gain for any given level of charge abstracted from the rear MCP of the chevron. That plates of the same geometry and glass type have different gain decay characteristics had earlier been noted by Sandel *et al.* (1977) and by Rees *et al.* (1980). That the 120:1 MCPs used in the lifetest of Whiteley *et al.* (1984) exhibit rather more rapid gain decay than any of those used in the Malina and Coburn study may be indicative of a real dependence of lifetime on channel length-to-diameter ratio.

Fig. 3.12b shows some very recent measurements made at Leicester using the 'Long Life' microchannel plates marketed by Galileo Electro-Optics (Cortez and Laprade, 1982). The new glass used in these night vision ($D = 10\,\mu m$, $L/D = 40:1$) MCPs has been specifically formulated to alleviate the problem of gain decay in image intensifier tubes. The gain in photon counting mode of two different 'Long Life' chevron pairs (fig. 3.12b) only falls to about 50% its original value even after charge levels of 0.1 C/cm² active area have been abstracted.

Longer life appears, in fact, to be a current preoccupation of all MCP manufacturers. Hamamatsu Photonics has also recently changed its MCP production process (curves i and ii of fig. 3.12b), while lifetime measurements made on the most recent night vision ($D = 12.5\,\mu m$, $L/D = 40:1$) MCPs produced by yet a third manufacturer, Mullard Ltd., closely correspond to those observed with declared longer-life plates.

Taking the results of figs. 3.12a and b together, one might reasonably conclude that not only is gain lifetime a rather uncertain quantity, it is a quantity liable to change from plate batch to plate batch as manufacturers

Fig. 3.12. Relative MCP gain as a function of charge abstracted per unit plate
area. (a) Lifetest data of Malina and Coburn (1984) (circles: Galileo MCPs;
squares: Mullard MCPs; crosses: Varian MCPs) and Whiteley *et al.* (1984)
(diamonds: Mullard MCPs). Full curve: lifetest on Hamamatsu MCPs ('old'
glass process: Matsuura *et al.*, 1985). Absolute initial gains for the Malina and
Coburn chevron detectors lay in the 0.53–0.94 pC range. The initial gain in the
chevron detector of Whiteley *et al.* was 3.2 pC per pulse. (b) Gain decay
measurements on Galileo Long Life MCPs (circles) and new Mullard night
vision MCPs (squares: Fraser *et al.*, 1988b) compared with lifetests of Hamamatsu MCPs made by both 'old' (curve ii) and 'new' (curve i) glass processes
(Matsuura *et al.*, 1985).

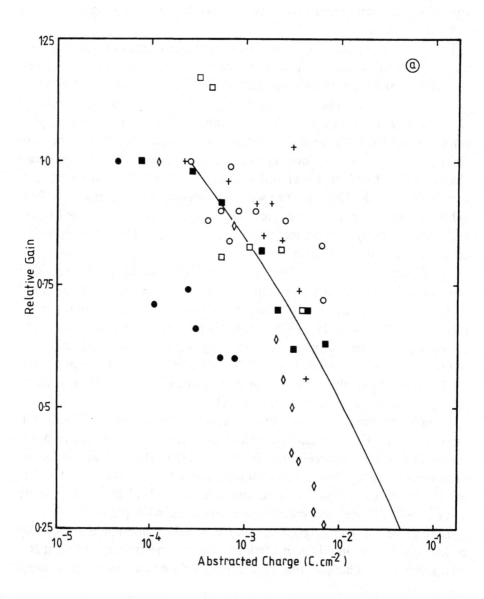

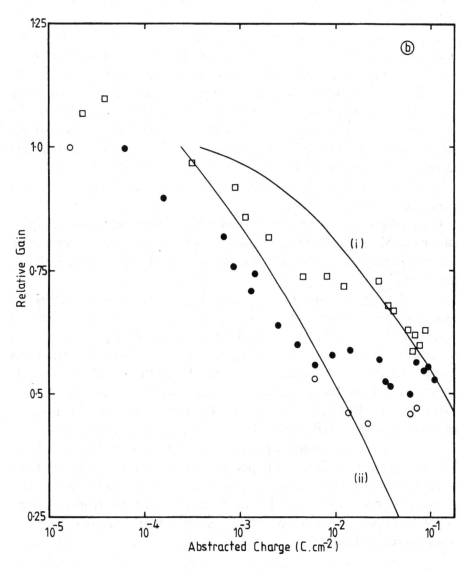

refine their processing techniques. Nor do lifetests such as these, conducted at constant bias voltage, tell the complete story. The useful lifetime of any MCP detector can be extended by several orders of magnitude beyond the levels indicated in figs. 3.12a and b simply by progressively raising the plate bias voltages to compensate for the ongoing fall in gain (Timothy, 1981). The gain of the chevron detector of Whiteley *et al.* (1984) was restored, at the end of the charge abstraction period, to its original level by a modest 250 V ($\sim$8%) increase in total applied voltage. Most experimental groups have their own 'burn-in' procedure for conditioning the gain of MCP detectors.

Even before the voltage increase some 4×10^9 counts/cm^2 had been accumulated by the detector in the 'worst case' lifetest of Whiteley *et al.* (1984). Such a dose corresponds to an average count rate of ~ 100 counts/cm^2 s for one year, a flux and duration typical of satellite-borne X-ray and XUV (Malina and Coburn, 1984) astronomy experiments. Thus, the provision in any flight detector of an adjustable high-voltage power supply effectively removes the possibility of experiment failure through MCP gain fatigue. The Einstein HRI microchannel plate detector (section 3.4.2) required only one voltage increment during its two years of in-orbit operation.

Gain fatigue is a more serious problem in some ground-based applications of MCPs (electron spectroscopy, synchotron research: see section 3.5) where particle or photon fluxes may be much higher than in X-ray astronomy.

Regarding the physics of gain fatigue, gain degradation curves are explicable on the assumption that an 'electron source population' is progressively removed from the active layers of the MCP glass by electron bombardment (Sandel *et al.*, 1977). According to some measurements (Authinarayanan and Dudding, 1976; Hill, 1976), this population consists of K$^+$ ions. It has been suggested that incorporation of less mobile Cs$^+$ ions into the MCP glass as a partial K$^+$ replacement would therefore eliminate gain decay (Bateman, 1979). The potassium migration theory, however, cannot be a complete picture. The glass from which Galileo Long Life MCPs are manufactured is potassium free (section 3.3.6), yet these plates still exhibit some gain reduction as charge is abstracted (fig. 3.12b). Alternative mechanisms may be the build up of an 'electron poisoning population', such as carbon, or simply the desorption of residual hydrogen from the reduced lead glass channel surfaces.

3.3.2.3 *Gain depression at high output count rates*

Charge depleted from the channel wall during the passage of a pulse must be replenished by the conduction current flowing between the MCP faces. By considering a 'lumped' *RC* circuit whose components are the channel resistance and an effective channel capacitance, it has been estimated (Wiza, 1979; Eberhardt, 1981) that charge replenishment should occur with a time constant $\tau = 10^{-2}$ s. Treatment of the MCP as a distributed impedance (Gatti *et al.*, 1983) indicates that the time for complete recharging exceeds 5τ.

On this basis we would expect that if a single channel were repeatedly excited at a frequency greater than $\sim 20\,\text{Hz}$, its gain would fall, since the steady state channel electric field would never be re-established. An output count rate of 12.5 counts/s per channel indeed produces a fractional gain reduction of 25% in a single curved channel MCP (Timothy, 1981). Pairs of conventional MCPs, however, operating at higher gains ($> 10^7$, compared with 1.2×10^6) can exhibit an equivalent fractional gain reduction at count rates of only 0.01 counts/s per channel (see figs. 3.13a, b). Somewhat *ad hoc* assumptions as to the magnitude of the lateral capacitance between the group of emitting channels and their quiescent neighbours must be made in order to extend the *RC* model to multistage detectors (Eberhardt, 1981). Siegmund *et al.* (1985) report measurements of 'gain droop' in Z-plate stacks. Pearson *et al.* (1988) have recently shown that MCP count rate characteristics depend on the fraction of the plate area illuminated.

An alternative way of analysing gain depression is to relate the magnitude of the pulse current delivered by the MCP detector (output count rate times charge gain) to the magnitude of the conduction current (MCP bias voltage divided by the resistance of the active area) actually available to replace the abstracted charge. The conduction current per channel in conventional 'high resistivity' MCPs is usually of order 10^{-11}–$10^{-12}\,\text{A}$.

Consideration of this charge supply limit (Loty, 1971) provides a consistent qualitative explanation of all the available gain-versus-count rate curves. The maximum pulse current which can be abstracted is found empirically to be between 10 and 30% of the available conduction current for single curved channel MCPs (Timothy, 1981), chevrons (Parkes and Gott, 1971; fig. 3.13a) and Z-plates (Siegmund *et al.*, 1985). Thus, the lower the charge per pulse and the lower the MCP resistance, the higher the count rates which can be accommodated. The former factor favours single curved channel MCPs; for imaging X-ray and XUV astronomy, however, the count rates accessible with stacks of conventional high-resistivity MCPs are usually sufficient.

MCPs of low resistance ($\sim 50\,\text{M}\Omega$) can now be produced by adjusting the final stages of glass processing (fig. 3.3). Conduction in MCPs is a surface phenomenon; the surface layers of channel plate glass have a negative temperature coefficient of resistance (section 3.2). Low-resistance plates, drawing relatively large wall currents, are thus in principle susceptible to 'thermal runaway' and self destruction due to excessive ohmic heating (Soul, 1971; Wiza, 1979; Timothy, 1981). In practice, the presence of an appropriate shunt resistor, in parallel with the MCP, in the detector high-voltage power supply completely removes this danger.

Fig. 3.13. Variation in chevron output pulse height distribution with output count rate. (a) Modal (peak) gain G_c. (b) Pulse height fwhm $\Delta G_c/G_c$. One output count/s is equivalent here to 0.05 counts/mm²s or 10^{-5} counts/channel·s. The fall in peak gain and the broadening of fwhm are caused by the limited conduction current in the rear MCP of the chevron, resistance 1×10^9 Ω. The presence of an insulating CsI layer at the input face of the front MCP has no effect on the chevron count rate characteristic.

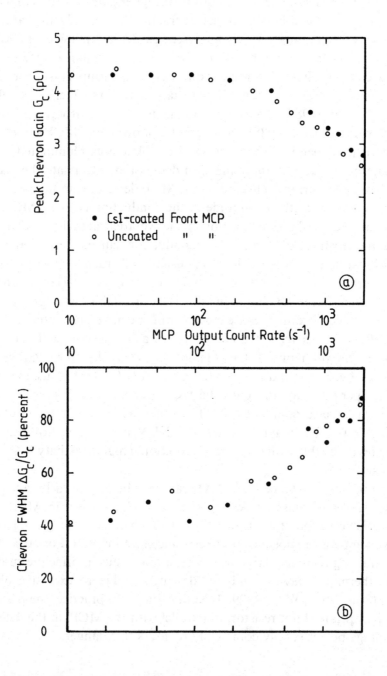

Bulk conductivity (hence, high conduction current) MCPs have several times been suggested (Washington, 1973; Gatti *et al.*, 1983), but never yet implemented, as high count rate devices.

3.3.3 *Quantum detection efficiency*

The soft X-ray and XUV quantum detection efficiencies of 'bare' MCPs (section 3.3.1) are rather low, typically in the 1–10% range (Parkes *et al.*, 1970; Timothy and Bybee, 1975; Kellogg *et al.*, 1976; Bjorkholm *et al.*, 1977; Bowyer *et al.*, 1981b; Fraser, 1982). They vary strongly with both photon energy (decreasing, in general, with increasing E) and angle of incidence θ. An efficiency-versus-angle curve has pronounced peaks at θ values correlated with the critical angle of X-ray reflection from the glass, and tends to zero for both normal ($\theta = 0°$) and grazing incidence to the MCP front surface. Efficiency measurements made on bare plates have been successfully compared with the predictions of detailed theoretical models (Bjorkholm *et al.*, 1977; Fraser, 1982).

In order to enhance the sensitivity of channel plates to soft X-rays, a material of high photoelectric yield may be deposited onto the MCP front surface and channel walls. Magnesium fluoride, chosen mainly on the grounds of photoelectric stability (Smith and Pounds, 1968; Lapson and Timothy, 1973), has been in the past the material of choice in X-ray astronomy. The channel plates in the Einstein HRI (section 3.4.2; Henry *et al.*, 1977) and EXOSAT channel multiplier array (CMA) detectors both incorporated 4000 Å thick MgF_2 photocathodes, as did the SOLEX solar X-ray spectrometer/spectroheliograph on the P78-1 satellite (Eng and Landecker, 1981).

Since 1982, however, a number of researchers have independently shown that caesium iodide is a much superior deposition photocathode not only in the soft X-ray band but at wavelengths from 2 to 2000 Å (Fraser *et al.*, 1982, 1985; Martin and Bowyer, 1982; Fraser and Pearson, 1984; Barstow *et al.*, 1985; Siegmund *et al.*, 1986b; Carruthers, 1987). Figs. 3.14a–d indicate the dependence of the efficiency of CsI-coated MCPs on (i) wavelength, (ii) incident angle, (iii) coating thickness, and (iv) exposure to a humid atmosphere. The production of CsI deposition photocathodes is described in detail by Fraser *et al.* (1982, 1984) and by Siegmund *et al.* (1986b). Coating stability at X-ray wavelengths is described by Saloman *et al.* (1980), by Premaratne *et al.* (1983) and by Whiteley *et al.* (1984). CsI, although hygroscopic, is rather stable if the MCP substrate is thoroughly baked out before deposition and if exposure to very humid atmospheres (RH > 50%) is excluded thereafter.

Fig. 3.14. Quantum detection efficiency of CsI-coated MCPs. (a) 2.1 Å efficiency as a function of incident angle and CsI layer thickness. Curve a: bare MCP. Curves b, c, d: 7000, 14 000 and 28 000 Å CsI layers deposited on MCP front surface, corresponding to stated thicknesses on channel walls (Fraser *et al.*, 1985). Vertical broken lines show expected angles of incidence from AXAF mirror cone (section 3.4.2). (b) 8.3 Å efficiency as a function of incidence angle (Fraser, 1984). Open circles: 14 000 Å CsI, deposited at coating angle $\hat{\alpha}_0 = 4°$ to channel axes. θ_c is the critical angle of reflection from both CsI and the underlying lead glass, corresponding to the angle of maximum efficiency. Filled circles: bare MCP. Curve: theoretical efficiency. (c) Peak efficiency as a function of wavelength for a variety of MCP storage conditions (Simons *et al.*, 1987). MCPs 1 and 2 were handled after coating in dry nitrogen; the latter MCP was subjected to a series of degradation tests, involving exposures to known humidities for known times. MCP3 was coated with 14 000 Å CsI and sealed under vacuum. These measurements were made as part of the Leiden programme to develop microchannel plate readout for an imaging GSPC (section 2.5.2). The broken curve represents the bare plate data of Martin and Bowyer (1982). (*Courtesy D. G. Simons.*) (d) Total (Q_T), front surface (Q_S) and open area (Q_C) efficiency contributions as functions of XUV wavelength (Barstow *et al.*, 1985). Incident angle $\theta = 30°$. The peak at 120 Å is due to the 4d–4f absorption feature in CsI, and has subsequently been confirmed by Siegmund *et al.* (1986b). (*Courtesy M. A. Barstow.*)

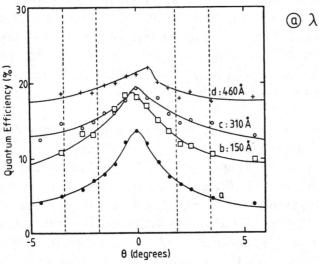

Fig. 3.14. (Continued)

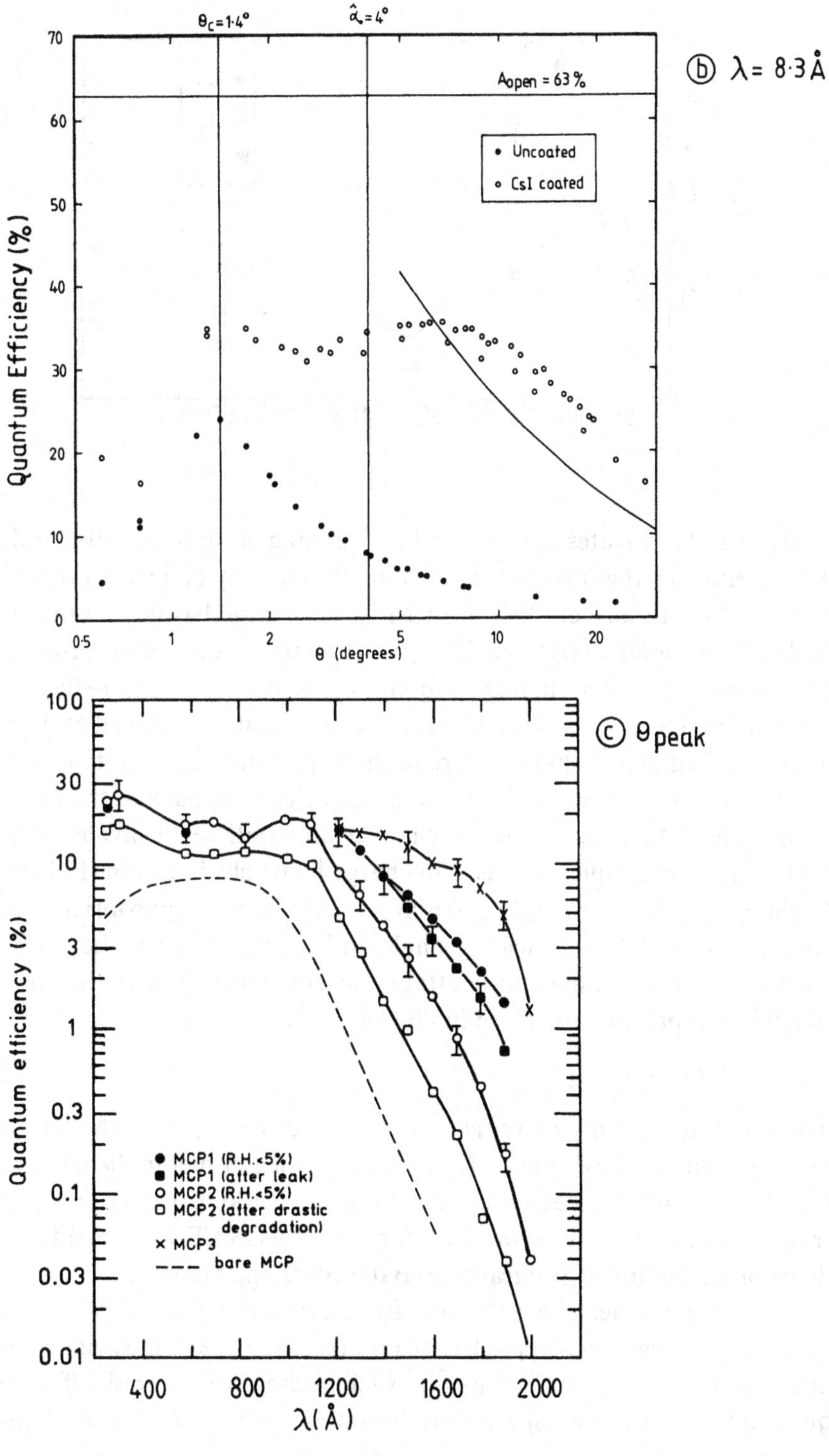

(b) $\lambda = 8\cdot3\,\text{Å}$

(c) θ_{peak}

Fig. 3.14. (Continued)

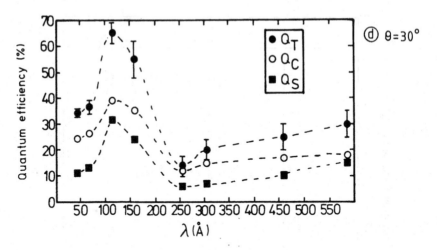

Fig. 3.14d illustrates the use of a field defining mesh or 'repeller grid' (fig. 3.8) to enhance the quantum detection efficiency by counting events from the MCP front surface. When an MCP is operated with a conventional high-voltage polarity (section 3.2), photoelectrons released by X-ray inter-actions with the inter-channel web are lost and the detection efficiency of the front surface, Q_S, is zero. A mesh of transparency T_0 placed at the MCP input and biased to a slightly more negative potential than that of the MCP front surface can, however, be used to repel electrons back to their plane of origin where they may enter a channel and initiate an avalanche. The repeller grid technique, first used in charged particle detection (Polaert and Rodiere, 1974; Panitz and Foesch, 1976), is not appropriate for all combinations of X-ray incident angle and energy. If Q_C is the open area efficiency and Q_S the maximum attainable front surface yield, then in order to achieve a practical increase in efficiency one requires

$$T_0(Q_C + Q_S) > Q_C$$

This condition is only marginally satisfied, for example, in the bare plate measurements of Taylor *et al.* (1983b), but is clearly met in the large-angle, long-wavelength CsI-coated plate data of fig. 3.14d, obtained with a 90% transmissive stainless steel mesh as part of the ROSAT Wide Field Camera detector development programme (Barstow *et al.*, 1985).

Since a photoelectron collected by the repeller grid may initiate an avalanche several channel pitches distant from the site of the original X-ray interaction, there is, in the use of such techniques, a tradeoff between quantum efficiency and spatial resolution (Taylor *et al.*, 1983b; Fraser *et al.*, 1984: section 3.3.5). For best imaging performance, large values of collecting field (>200 V/mm) are required to limit the transverse electron range. Such fields inevitably reduce the chances of electrons originating on the MCP front surface entering any neighbouring channel.

The use of 'funneled' MCPs (section 3.2) would in this respect appear a better means of increasing the effective area of channel plate detectors. No X-ray measurements have so far been reported on such devices. Considering the X-ray interaction geometry, however, leads one to conclude that the efficiency enhancement offered by funneling would strongly depend on the relationship of the half-angle of the channel entrance cone and the angle of X-ray incidence.

Another alternative is to use a separate transmission photocathode mounted on an X-ray transparent substrate in front of the MCP detector. Normal density transmission photocathodes (deposited in high vacuum) have few advantages over deposited reflection photocathodes at soft X-ray wavelengths (Henke *et al.*, 1981; Fraser, 1985). So-called 'fluffy' CsI layers with about 5% bulk density, deposited in a few Torr atmosphere of argon, have, however, aroused intermittent interest ever since the measurements of Bateman and Apsimon (1979) and Bateman *et al.* (1981) indicated 5.9 keV efficiencies in excess of 50% (cf. fig. 3.14a). Low-density alkali halide layers, with their electron emission enhanced by the application of strong (~ 1 kV/mm) electric fields, were first used as charged particle detectors (Edgecumbe and Garwin, 1966). A recent thorough study (Kowalski *et al.*, 1986) has shown that while fluffy CsI cathodes are somewhat less reproducible than bulk density coatings, they yield, at all wavelengths in the band 1.5–400 Å, broadly similar efficiencies to those displayed in figs. 3.14a–d.

The most recent photocathode research at Leicester and at the Space Science Laboratory, Berkeley, has concerned materials still more efficient than CsI in selected wavelength ranges. CsBr in the 20–100 Å band (Fraser *et al.*, 1987a) and KBr in parts of the 44–1500 Å band (Siegmund *et al.*, 1987, 1988a) have been shown to provide further enhancement factors of order two in coated plate efficiencies.

As noted in section 3.3.1, the soft X-ray detection efficiency of an MCP is independent of channel diameter D and pitch p. As X-ray energies (and photoelectron ranges) increase, the X-ray beam may penetrate a septum and excite electrons into more than one channel. The detection process passes from a 'single channel' mode to a 'bulk absorber' mode, in which thick MCPs with thin channel walls (large L, small $p\text{-}D$) provide the best detection efficiencies (Bateman, 1977a; Gould *et al.*, 1977; Fraser, 1982;

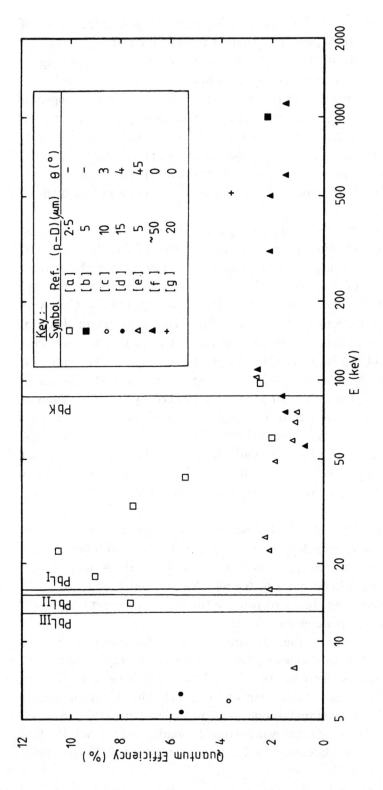

Fig. 3.15. Lead glass MCP hard X-ray and γ-ray quantum efficiencies. Note: (i) the superiority of thin-walled channels for E <50 keV;
(ii) the influence of the lead K edge (rightmost vertical line); (iii) the
weak dependence of efficiency on incident angle when the MCP
operates as a bulk absorber. Compton scattering is important in
determining the efficiency only above about 100 keV. *References*: (a)
Bateman (1977a); (b) Timothy and Bybee (1979); (c) Parkes *et al.*
(1970); (d) Eng and Landecker (1981); (e) Dolan and Chang (1977);
(f) Adams (1966); (g) McKee *et al.* (1985).

McKee *et al.*, 1985). Measured hard X-ray and γ-ray efficiencies are shown in fig. 3.15. The efficiency is remarkably constant over a very broad energy range. This constancy is also apparent in the 28–140 keV relative efficiency data of Gould *et al.* (1977). The efficiency of 511 keV positron annihilation gammas may be as high as 9% in MCPs of fully optimised geometry (McKee *et al.*, 1985). Thick, large-area MCPs may in future find application as detectors of unsurpassed spatial and temporal resolution (but still relatively low efficiency: see section 5.2) in hard X-ray astronomy, positron emission tomography and electron momentum spectroscopy (section 3.5).

3.3.4 *Energy resolution*
Until recently, this section might reasonably have consisted of the single word 'none'. CsI-coated chevron MCP detectors of high quantum detection efficiency can, however, now be made to display a limited degree of soft X-ray energy resolution provided the front MCP is operated with a bias voltage V_F less than the saturation level $(V_0)_s$ (section 3.3.2.1). The energy dependence of the number of photoelectrons released from a channel coating of CsI at grazing incidence is then translated, semi-proportionally (eq. 3.1), into a variation of chevron gain G_c with X-ray energy (Fraser and Pearson, 1984). Fig. 3.16 shows the best energy

Fig. 3.16. Overlaid C K (0.28 keV) and Si K (1.74 keV) MCP pulse height distributions, each containing 70 000 counts. The ratio of the peak gains is 1.8:1. Some 37% of the Si K counts lie in the overlap of the two distributions.

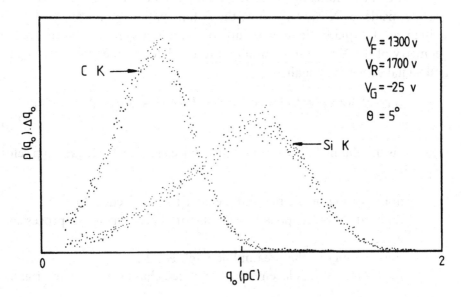

discrimination achieved to date, with the aid of a retarding inter-plate electric field (Fraser *et al.*, 1985).

While it is not really appropriate to assign $\Delta E/E$ values to channel plate detectors on the basis of such data, 'two-colour' photometry should be possible in future detectors such as the AXAF HRC (section 3.4.2), without recourse to the succession of bandpass filters necessary on the Einstein and EXOSAT missions (section 3.3.6). Any measure of energy resolution is useful in imaging X-ray astronomy in discriminating against the soft X-ray diffuse background (B_d in eq. 1.1). Siegmund *et al.* (1985) have shown that the different reflection probabilities, and hence different interaction depths within the channel (section 3.3.2.1), of X-ray and UV photons lead to differences in Z-plate gain. MCP energy resolution typified by fig. 3.16 could also be used to reduce the effects of high-energy mirror scatter on image quality.

Improvements in MCP spectral response will depend on the development of plates with higher secondary electron yields and hence 'more proportional' avalanche characteristics. Ultimately, the achievable spectral resolution is set by the statistics of photoemission from the coating material (Fraser and Pearson, 1984).

A phenomenon analogous to the dependence of MCP gain on X-ray energy – its dependence on positive ion charge state – has been reported by Tagaki *et al.* (1983).

3.3.5   *Spatial resolution*

For any multistage microchannel plate detector at the focus of a grazing-incidence X-ray telescope, the fwhm spatial resolution Δx can be written as the quadratic sum of up to eight terms, four relating to the geometry of the X-ray interaction and four related to the readout element and signal processing chain:

$$\Delta x = [2.36\{\Delta x_a^2 + \Delta x_{d'}^2 + \Delta x_e^2 + w_f \Delta x_f^2 + \Delta x_{g1}^2 + \Delta x_{g2}^2 + \Delta x_p^2\} + D^2]^{\frac{1}{2}}$$

(3.2)

where D is the channel diameter and the suffices a, d′, e, f, g1, g2 and p refer respectively, to:

noise arising in the readout element itself (cf. eq. 2.15);
digitisation of the position signals (or equivalent signal processing error);
noise arising in the preamplifiers (cf. eq. 2.15);
the transverse displacement of activated channels from the original

positions of X-ray interaction on the interchannel web (section 3.3.3), where w_f is the fraction of all events originating on the web; jitter in the charge cloud centroid arising in the inter-plate gap(s); jitter in the charge cloud centroid arising in the MCP-readout gap; partition noise (for a charge-sharing readout: section 2.4.1.1).

Since the functions of detection/amplification and of position encoding are separable in MCP detectors, a wide range of detector geometries has evolved, each with its resolution dominated by a different term in the series. MCP encoding options, some of which have already been discussed in a proportional counter context (section 2.4.1.2), are described more fully in section 3.4.1. In all cases, however, the fundamental limit to resolution is the channel diameter. Individual 25 μm microchannels have been imaged several times, usually over very small (~ 1 mm) fields-of-view (Lampton and Malina, 1976; Lapington *et al.*, 1987). As noted in the introduction to this chapter, manufacturers are now producing plates with pore diameters much smaller than 25 μm.

Comparing eq. 3.2 with eq. 2.11, its equivalent for a focal plane proportional counter, we note that there are no X-ray penetration, diffusion or photoelectron range terms in the former expression. This reflects an important advantage of all solid state imaging detectors over their gas-filled competitors. Obviously, as X-ray energies increase, both $1/e$ absorption lengths and photoelectron ranges must become comparable with the average septal thickness $p - 2D/\pi$ of the channel plate (Fraser, 1982). For all current astronomical applications, however, limited by the response of grazing-incidence optics to energies below 10 keV, MCP detectors are essentially 'parallax-free'.

Bateman (1977a) discusses the expected hard X-ray (10–140 keV) spatial resolution of MCP detectors. Even with 511 keV positron annihilation gammas, fwhm resolutions of 2.5 mm should still be attainable using thick MCPs (McKee *et al.*, 1985).

Fwhm position resolution, finally, should not be regarded as a complete descriptor of imaging power. As noted by Siegmund *et al.* (1986a), the centroid of the point spread function of any detector with continuous, digitally oversampled position signals can be determined to a factor $N^{\frac{1}{2}}$ better than the detector resolution, where N is the number of counts in the feature. This property – labelled positional sensitivity by some authors (cf section 2.4.1.1) – is important in MCP detectors for grating spectroscopy, in determining the wavelengths and profiles of spectral lines.

3.3.6   *Dark noise*

The internal background count rate in the current generation of MCP detectors is usually uniformly distributed over the plate area at a density ~ 0.2 counts/cm^{2}s at sea level. This 'universal' MCP count rate, which is independent of X-ray bandwidth, is between three and four orders of magnitude higher than the comparable figure for a position-sensitive proportional counter such as the Einstein IPC (section 2.4.2), assuming a 1 keV bandwidth centred on 1 keV. This remarkable difference between detector types reflects, on the one hand, the sophistication of the background rejection techniques now implemented in proportional counters and, on the other, long-standing ignorance of the nature of the noise sources in MCPs.

MCP noise pulse height distributions are exponential in form (fig. 3.17), indicating a source uniformly distributed throughout the plate volume (section 3.3.2.1). Non-thermal noise sources considered in the literature have included field emission from channel defects (Henry *et al.*, 1977) and cosmic rays. Recently, a study of MCP dark noise conducted at Leicester (Fraser *et al.*, 1987b) has identified internal radioactivity as the principal ($>90\%$) noise source in well-outgassed, blemish-free MCPs, operated at pressures below 10^{-6} mbar. This conclusion has since been confirmed by studies at Berkeley (Siegmund *et al.*, 1988b).

Most channel plate glasses contain $\sim 5\%$ by weight of potassium for mechanical (ease of drawing of fibre, fig. 3.3) and electrical (to enhance the secondary electron yield of the finished multiplier, Hill, 1976) reasons. The long-lived (half-life 1.28×10^9 years) beta emitter ^{40}K is present as 0.0118% of naturally occurring potassium. Monte Carlo calculations of the effects of the ^{40}K emission in Mullard MCPs accurately account for both the absolute noise count rates and pulse height distribution shapes (fig. 3.17) observed experimentally. Given that the elimination of radioactive potassium is a major concern in the manufacture of alkali iodide scintillation counters (Chapter 5), it is perhaps rather surprising that the role of internal radioactivity in MCPs was not identified earlier. Glasses used by US

Fig. 3.17. (a) MCP detector signal pulse height distribution. (b) Dark noise pulse height distribution. (c) Signal-to-noise ratio, S, as a function of lower-level discriminator setting, for the distributions (a) and (b). The common abscissa is the output charge, q, expressed in terms of the modal X-ray gain ($G_c = 8.4$ pC). Signal and noise are poorly separated. Optimum noise discrimination (maximum S) is obtained when all events (including 51% of the X-rays) below $0.9\,G_c$ are rejected. The lower discriminator settings which appeal to the eye on inspection of the peaked signal distribution result in S values a factor of two lower (Fraser *et al.*, 1987b).

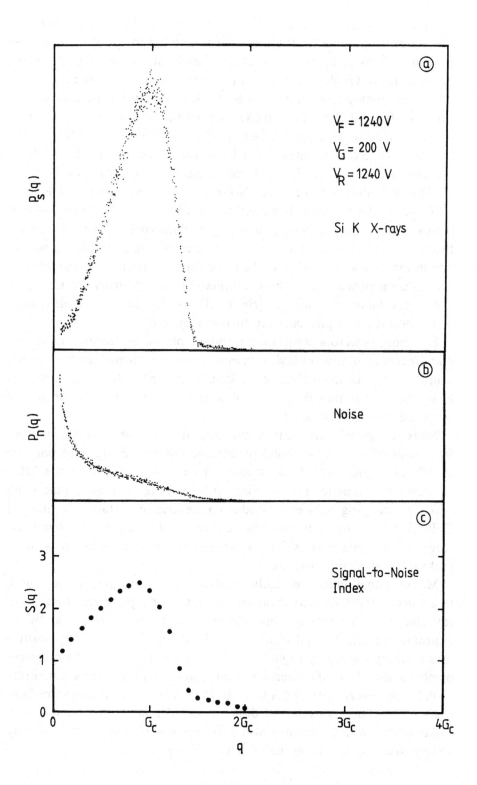

manufacturers (Table 3.1) commonly contain, in addition to potassium, a few per cent by weight of rubidium. ^{87}Rb, another long-lived beta emitter present as 28% of all rubidium atoms, has been shown to be the major noise source in Galileo's Long Life plates (Fraser *et al.*, 1988b).

Current attempts in both the US and UK to develop potassium- and rubidium-free MCP cladding glasses, should they prove successful, offer a ten-fold reduction in ground-based detector noise, leaving cosmic rays at a level of ~ 0.015 counts/cm^2 s as the dominant source (Fraser *et al.*, 1987b). It is expected, moreover, that such reductions would carry forward into the satellite radiation environment. Noise levels in the Einstein HRI and EXOSAT CMA detectors increased in orbit by factors less than two over their sea level values. Elimination of the MCP internal radioactivity should therefore result in at least a two-fold enhancement in instrumental sensitivity in the 1–10 keV band, even before external particle rejection schemes such as coincidence with a plastic scintillator shield (Murray and Chappell, 1985) or vetoing extended particle 'tracks' of activated channels (Fraser *et al.*, 1987b) are implemented to further reduce B_i.

For those detectors containing standard plates, meanwhile, an immediate (experimentally verified) consequence of the identification of noise sources is that the front plate of the stack should be as thin (i.e. contain as little radioactive material) as is consistent with the requirements of saturated gain (section 3.3.2.1).

Noise 'hotspots' – the local brightening of the noise image to produce false 'sources' – are a perennial problem in MCP imaging. Hotspots are usually associated with field emission from dust particles on the MCP surfaces or from plate imperfections, with poor high-voltage contacts and with local trapping of gases evolved from the channel surfaces (Malina and Coburn, 1984). Hotspots can thus be eliminated by clean plate handling, careful plate mounting (fig. 3.7a) and the provision of adequate outgassing paths in the detector structure.

Microchannel plates are highly sensitive to electrons, positive ions and, to a much lesser degree, to ultraviolet light. Channel plate detectors for X-ray and XUV astronomy must therefore be preceded (fig. 3.8) by an appropriate earthed shield with the dual function of (i) preventing positive ions reaching the highly negative MCP input, and (ii) suppressing instrumental sensitivity to the intense UV emission from the earth's geocorona (notably the 304 Å HeII, 584 Å HeI and 1216 Å H Lyman α lines) and from hot stars. Using a selection of different shield materials, mounted on a rotating filter wheel, one may also perform broad-band spectroscopy by viewing the sky in turn through different X-ray bandpasses.

The quantum detection efficiency of bare MCPs and CEMs is a steeply falling function of wavelength beyond about 1200 Å (fig. 3.14c; Martin and Bowyer, 1982; Paresce, 1975), reaching the $10^{-7}\%$ level at 2500 Å. Bare (and MgF$_2$ coated) MCPs are thus in principle 'solar blind'. The addition

Fig. 3.18. ROSAT High Resolution Imager UV/ion shield. (a) X-ray transmission versus energy. (b) UV transmission versus wavelength. The locations of the brightest geocoronal emission lines are indicated (Pfeffermann *et al.*, 1987). (*Courtesy M. V. Zombeck.*)

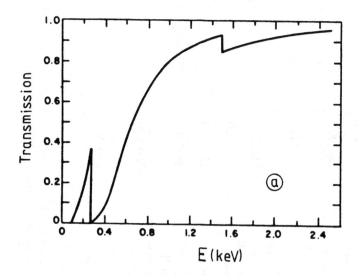

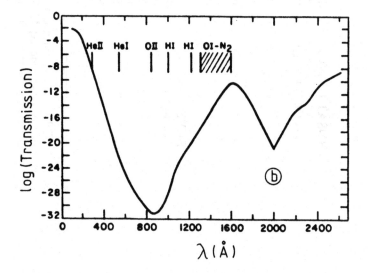

Table 3.2. *Ultraviolet light/positive ion shields for satellite-borne MCP X-ray detectors*

Detector	Filter	Bandpass (Å)	Reference
Einstein Obs.			Giacconi *et al.*
HRI (M)	~1 μm par N	<25[a]	(1979) (see Table 3.3)
	+ ~0.05 μm Al		
P78-1 SOLEX (M)	Polypropylene + Al	<25	Eng and Landecker (1981)
EXOSAT CMA (M)	0.3 μm Lexan	44–300	de Korte *et al.* (1981b)
	0.1 μm par N	160–600	
	+ 0.1 μm Al		
	1 μm B	68–100	
ROSAT HRI (C)	1 μm par N	<25[a]	Pfeffermann *et al.* (1987)
	+ 0.085 μm Al		
ROSAT WFC (C)	0.2 μm Lexan	60–140	Barstow and Pounds (1987)
	+ 0.2 μm C		
	+ 0.1 μm B (S, P)		
	0.2 μm Lexan	112–200	
	+ 0.3 μm Be (S, P)		
	0.2 μm Lexan	150–220	
	+ 0.2 μm Al (P)		
	0.2 μm Sn	530–720	
	+ 0.2 μm Al (P)		
EUVE survey	0.2 μm Lexan	68–250	Vallerga *et al.* (1986)
detectors (C)	+ 0.12 μm B		
	0.4 μm Al + 0.06 μm C	160–350	
	0.02 μm Ti +	380–600	
	0.2 μm Sb +		
	0.02 μm Ti		
	0.06 μm In +	500–850	
	0.015 μm SiO +		
	0.2 μm Sn		

[a] Filters which feature a narrow transmission band associated with the C K absorption edge.
par N=parylene N; M=MgF$_2$-coated microchannel plate; C=CsI-coated microchannel plate;
S=survey filter; P=pointing mode filter.

of a CsI photocathode (or, to a lesser extent, KBr or CsBr photocathode) to the detector pushes the UV sensitivity out beyond 2000 Å (fig. 3.18c), but preserves the exponential fall-off in efficiency with increasing wavelength. UV filter design is therefore somewhat more demanding for CsI-coated MCPs (a thicker aluminium layer may in some cases be all that is required; Table 3.2), but it remains straightforward compared to the problems encountered with the silicon-based XUV detectors now being proposed as imaging rivals to MCPs in the 100–1000 Å band (section 4.6.2.3). CCDs and diode arrays inevitably have additional sensitivity at optical wavelengths where microchannel plates are totally blind.

Just as in the case of proportional counter windows, thin metal foils or metallised plastics provide the appropriate X-ray transmission and UV absorption characteristics. Filters are now constructed on their support meshes as multilayer sandwiches, in order to eliminate transmission through pinholes (Barstow *et al.*, 1987). Table 3.2 describes the UV/ion shields used, or planned for use, in a number of satellite-borne microchannel plate X-ray detectors. Figs. 3.18a and b show the expected X-ray and UV transmissions of the Al/parylene N (par N) shield designed for the CsI-coated ROSAT High Resolution Imager (Pfeffermann *et al.*, 1987).

Thin plastic filters in low earth orbit may be subject to chemical erosion, mass loss and mechanical failure if attacked by atomic oxygen (Albridge *et al.*, 1987). Currently, this is a major concern for planned experiments such as the ROSAT Wide Field Camera, where boron overlayers have added to some filter designs to prevent erosion (Table 3.2).

3.3.7 *Temporal resolution*

As with proportional counters (section 2.3.4), the time resolution of satellite-borne MCP detectors is usually determined by the spacecraft telemetry chain. It is important to note, however, that MCPs are intrinsically very fast devices. The pulse transit time through the intense (~ 1 kV/mm) electric field of a typical plate is of order 10^{-10} s, a feature which has been exploited in a number of fast photomultiplier tube designs (Boutot *et al.*, 1977; Hocker *et al.*, 1979; Lo and Leskovar, 1981; Oba and Rehak, 1981; Moszynski *et al.*, 1983).

The transit time τ_{MCP} for a single plate of length-to-diameter ratio L/D operating in linear mode with an applied voltage V_0 is given by the expression (Adams and Manley, 1966)

$$\tau_{MCP} = n\tau = \{4(L/D)^2(V/V_0)\}\{D(m/2eV)^{\frac{1}{2}}\} \tag{3.3}$$

Here, n is the average number of electron collisions with the channel wall and τ is the time between collisions. e/m is the charge-to-mass ratio of the electron and eV is the average energy of an electron emitted perpendicularly from the channel wall. According to Eberhardt (1981), the transit time spread, the fwhm of the distribution of transit times, is simply

$$\Delta\tau_{MCP} = n^{\frac{1}{2}}\tau \qquad\qquad (3.4)$$

For a single 40:1, 12.5 µm channel diameter MCP, subjected to a bias of 1000 V, assuming $V = 1$ V (Hill, 1976), eqs. (3.3) and (3.4) predict

$$\tau_{MCP} = 135\,\text{ps}; \ \Delta\tau_{MCP} = 53\,\text{ps}$$

Such analytical estimates are borne out both by Monte Carlo models of the avalanche process (Guest, 1971; Ito *et al.*, 1984) and by experiment. Inspection of equation (3.3) reveals that MCP time resolution scales directly with channel diameter; current attempts to manufacture plates with 2 µm and 4 µm channels (section 3.1) are directed towards the production of ultra-fast photomultipliers.

A second consequence of the intense electrical fields and small dimensions of the microchannel plate is its relative immunity to magnetic fields. This detector property is important in particle spectrometers. The maximum tolerable axial magnetic field (i.e. the maximum field component along the channel axis) can be simply estimated from the condition that the radius of gyration of an electron emitted perpendicularly from the channel wall equal the channel radius. One obtains:

$$B_{max} = (8Vm/e)^{\frac{1}{2}}/D$$

For $B > B_{max}$ (~ 0.54 T for 12.5 µm channels) electrons fail to strike the channel wall and the multiplier gain falls. The behaviour of two- and three-plate stacks in axial and perpendicular magnetic fields of up to 3 T has been described by Bateman *et al.* (1976), by Oba and Rehak (1981) and by Morenzoni *et al.* (1988).

While neither the intrinsic time resolution of microchannel plates nor their magnetic field immunity have yet been exploited in space astronomy, possible future applications (accurate photon arrival time-tagging, event discrimination at high count rates) have been identified, at least for the fast time response (Siegmund *et al.*, 1986a). Very recently, McMullan *et al.* (1987) have shown that fast timing ($\Delta\tau_{MCP} = 500$ ps) is possible even in detectors incorporating 'slow' (section 2.4.1.2) resistive anode readout, if the temporal information is recovered from the pulse induced on the MCP output electrode by the passage of the charge cloud. This ability to decouple imaging and timing may find important applications in X-ray astronomy.

3.3.8 *Polarisation sensitivity*

Researchers at the University of Windsor in Canada (McConkey *et al.*, 1982: Tomc *et al.*, 1984) have recently shown that bare MCPs and CEMs have a significant sensitivity to the linear polarisation state of XUV radiation, at wavelengths as short as 304 Å (fig. 3.19). The emphasis of their reports has been to caution that, in any experiment involving polarised

Fig. 3.19. Polarisation sensitivity of MCPs (Tomc *et al.*, 1984). The graphs show the measured variation of the modulation factor, M (section 1.3), with angle of photon incidence, θ, relative to the plate normal. M has its maximum value for all three wavelengths (He, 304 Å; N_2, 950–1000 Å; H_2, 1216 Å) when the incident beam is coaligned with the channel axes, which lie at an angle $\theta_B = 8°$ to the plate normal (fig. 3.8). (*Courtesy J. Tomc.*)

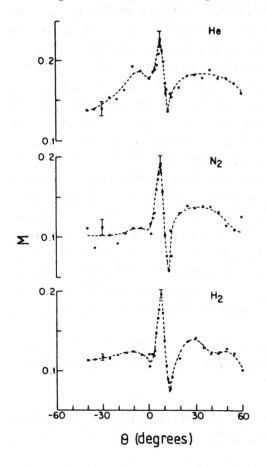

light, the response of the MCP may be a complicating factor. In an astrophysical context, this warning is relevant to future UV spectroscopy missions such as ESA's Lyman (Aschenbach *et al.*, 1985), in which MCPs will be required to detect 900–1200 Å radiation partially polarised by reflection from gratings. Presumably the preferential response of the MCP to p-polarised radiation results from differences in the reflection and absorption coefficients of the lead glass for the two planes of polarisation.

Viewing the data of fig. 3.19 in a more positive light, it implies that a channel plate polarimeter with a modulation factor (section 1.3) $M = 0.25$ and an efficiency of a few per cent could be constructed to operate at 304 Å. Given the wavelength independence of the effect exhibited in fig. 3.19, it is interesting to speculate how far the polarisation sensitivity of MCPs (or indeed planar photocathodes) may extend into the soft X-ray band.

3.4   Imaging

Small MCPs were the first detectors of high spatial resolution to be used in soft X-ray astronomy (section 1.4). Microchannel plates have now been displaced from that niche by cooled *charge coupled devices* (CCDs: section 4.6), devices of vastly superior energy resolution, and have acquired the 'low technology' label which X-ray astronomers automatically attach to detector concepts which have actually flown, or are about to fly (Table 3.2), in space. Developments discussed earlier in this chapter – high efficiency coatings, low-noise glass, large areas, fast timing – are, however, likely to ensure that MCPs remain the detectors of choice for a relatively large number of future X-ray, XUV (and UV) satellite missions, particularly those involving (transmission) grating readout and/or imaging in the 100–1200 Å 'open window' band. Such missions include ESA's Lyman (Aschenbach *et al.*, 1985) and the proposed SPECTROSAT follow-up to the West German Röntgensatellit (section 1.5). Since dark noise sources in MCPs are non-thermal (section 3.3.6), microchannel plates have the practical advantage over CCDs that they require no cooling. In the long run, the relatively low cost of large-area MCPs ($\sim £50/cm^2$ image area) may assume some importance.

This section describes the construction and operation of imaging microchannel plate detectors with reference to the following satellite-borne instruments:

(i) the Einstein Observatory High Resolution Imager (HRI: Kellogg *et al.*, 1976; Henry *et al.*, 1977; Giacconi *et al.*, 1979) and its descendants, the ROSAT HRI (Pfeffermann *et al.*, 1987) and the AXAF High Resolution Camera (HRC: Murray and Chappell, 1985);

(ii) the ROSAT Wide Field Camera detectors (Barstow *et al.*, 1985; Wells, 1985).

3.4.1 *Position encoding in microchannel plate detectors*

As noted in section 3.3.5, the family of MCP readout methods is a large one. Despite important differences in the way signals are sensed in gas detectors and MCP detectors (as slow-rising, extended induced charge distributions in the former case, and as fast, localised electron pulses in the latter) this family includes as a subset nearly all the proportional counter readout methods discussed in sections 2.4.1.1 and 2.4.1.2. Direct computation of the charge centroid from the signals on parallel conducting strips is perhaps the one gas detector method not yet implemented in a channel plate detector.

Capacitive charge division was one of the earliest MCP readout methods investigated for possible use in astronomy (Gott *et al.*, 1970; Morton and Parkes, 1973). Using a single MCP and a grid of parallel 100 μm diameter wires, set on 200 micron pitch and coupled by 100 pF capacitors, Gott *et al.* (1970) achieved fwhm resolutions of 75 μm over a 4 mm active area. The preamplifier term Δx_e (eq. 3.2) generally dominates in such systems. Δx_e is in turn determined by the preamplifier input capacitance. Thus, increasing the inter-wire capacitance values of the divider chain, relative to the stray capacitance to ground, would have had the effect of improving the integral non-linearity δ (section 2.4.1.1) of the encoder at the expense of its spatial resolution.

One-dimensional position encoding using MCPs and resistive strips was investigated by a number of research groups in the early 1970s. Parkes *et al.* (1974), working at Leicester, demonstrated resolutions of ~40 μm fwhm over a linear field of 18 mm, thus approaching, with the zero-cross timing method of section 2.4.1.1, to within a factor two of the channel diameter limit of their chevron detector (eq. 3.2). Sounding rocket UV spectrographs incorporating MCPs and charge-dividing resistive strips are described by Lawrence and Stone (1975) and Weiser *et al.* (1976). The partitioned *RC* line encoder (fig. 2.21c) developed for the Einstein and ROSAT HRIs is described in some detail in the following section.

True two-dimensional resistive anodes have been used in microchannel plate detectors for both the EXOSAT and ROSAT missions. Anode design for the EXOSAT CMA experiment was identical to that used in the parallel-plate avalanche chamber, the PSD, on the same satellite (section 2.4.3; fig. 2.20d). The curved resistive anodes designed for the ROSAT Wide Field Camera are described in detail in section 3.4.3.

At first sight, it may appear paradoxical that the 'distortionless' resistive anode of Gear's design (fig. 2.20e) has not figured in any satellite detector design. Gear anode noise levels, however, are somewhat larger (~ 3750 electrons rms equivalent noise charge; Lampton and Carlson, 1979) than those of an equivalent uniform resistive anode (1800 electrons rms; Barstow et al., 1985) due to the shunt effect of the boundary resistors. Hence, the Gear anode resolution limit Δx_a (eq. 3.2) is higher than might otherwise be achieved. Fixed image distortion can always be corrected; resolution once lost can never be regained. The work of Lampton and Carlson (1979) at the Space Sciences Laboratory, Berkeley, has, however, led to the commercial availability of MCP detectors incorporating Gear anode readout so that, of all the microchannel plate position encoders, the Gear design is certainly the most commonly used (Table 3.1).

It was also at Berkeley that the original work on wedge-and-strip (WS) anodes was carried out (Martin et al., 1981). Space Sciences Laboratory researchers have since developed microchannel plate detectors with WS readout for the EUVE mission (section 1.5). The imaging performance of these 50 mm diameter detectors is illustrated in fig. 3.20 (Siegmund et al., 1986a, c). Fwhm spatial resolutions as small as 80 µm have been measured in the EUVE development programme. Fig. 3.21a shows the dependence of the electronic (Δx_e) and partition noise (Δx_p) components of the three-element (fig. 2.22c) WS resolution on EUVE detector gain. At all practical gains ($< 5 \times 10^7$ electrons/photon) in detectors of this size, loading of the charge-sensitive preamplifiers by the inter-electrode capacitances of the anode dominates the overall resolution. Partition noise resolution, which varies inversely as the square root of the gain (section 2.4.1.1), would only dominate at gains approaching 10^8. Partition noise dominance (and, indeed, channel diameter dominance of the overall resolution) is observable with much smaller (~ 5 mm) anodes at much lower gains (fig. 3.21b). The erroneous claim by Schwarz (1985), that quantum partition noise

Fig. 3.20. Imaging performance of EUVE wedge-and-strip readout (Siegmund et al., 1986a, c), evaluated by the pinhole mask method (section 2.2.5). (a) 2500 Å image of 50 µm diameter pinhole array. Horizontal spacing = 2 mm. Vertical spacing = 4 mm. (b) Image linearity derived from (a). Position displacement is the difference between measured and true x-axis pinhole positions, expressed in channels of 12.5 µm width. (c) Spatial resolution derived from (a), as a function of x-axis position. (*Courtesy O. H. W. Siegmund.*)

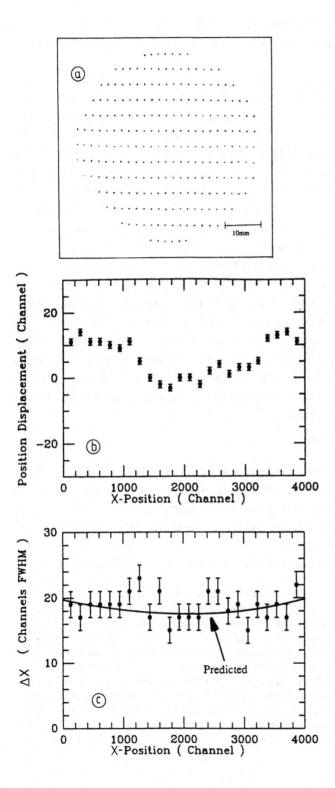

would not be observed in practical devices, was based on measurements involving capacitively induced signals rather than the actual collection of a number of electrons.

An increasing concern in MCP detector development is how to combine large image areas with spatial resolution at, or near, the channel diameter level. Using the two-amplifier-per-axis amplitude ratio methods discussed above, the limiting contribution in eq. 3.2 usually turns out to be Δx_d, the finite resolution of the signal processing chain. In a modern amplitude ratio circuit, a peak sample-and-hold unit and analogue-to-digital converter (ADC) form the basis of the 'divide' unit (fig. 2.17c). For an encoder length L, using an N-bit ADC in each of four signal channels implies that $\Delta x \sim L/2^{N-1}$, or about 50 μm using 12-bit fast ADCs to achieve reasonable count rates in a detector 10 cm square.

Partitioning the image axes (section 2.4.1.2) is one solution to this problem, implemented in the Einstein and ROSAT HRIs described in the next section. Another solution may be to use delay line encoding (section 2.4.1.2) where the electronic requirement is to sense the arrival time difference for pulses at the two ends of the delay line. Bateman *et al.* (1982) have used constant fraction discriminators with 6 ps precision to time the signals coupled into a 25 cm long, 1 ns/mm delay line by means of a 'fan-out' array of printed circuit board tracks, achieving 2.5 μm fwhm resolution in one dimension. That is:

$$\Delta x_d/L = 1 \times 10^{-4}$$

a performance which would require 14- or 15-bit ADCs in any amplitude ratio system. Overall detector resolutions $\Delta x/L = 1 - 6 \times 10^{-3}$ have recently been reported by Keller *et al.* (1987a), by Lampton *et al.* (1987) and by Sobottka and Williams (1988) using a variety of delay-line-based two-dimensional readout schemes.

Digital encoding schemes (section 2.4.1) provide a third 'escape' from the

Fig. 3.21. Fwhm spatial resolution contributions in MCP detectors with wedge-and-strip readout. (a) EUVE detector data (Siegmund *et al.*, 1986a, c). The crosses represent measurements made by capacitively coupling known charges to the input of the signal processing chain. Note the different slopes of the preamplifier and partition noise components of the resolution. In this data set, measured MCP detector resolutions (circles) retain a contribution from pinhole size. (*Courtesy O. H. W. Siegmund.*) (b) Small scale anode data (Lapington *et al.*, 1987). Detector resolutions (squares) here refer to the blurring of the image of a single 25 μm diameter curved microchannel. Note that the limiting fwhm spatial resolution deduced on the basis of a Gaussian point response function is actually less than the channel diameter D. (*Courtesy J. Lapington.* © *1987 IEEE.*)

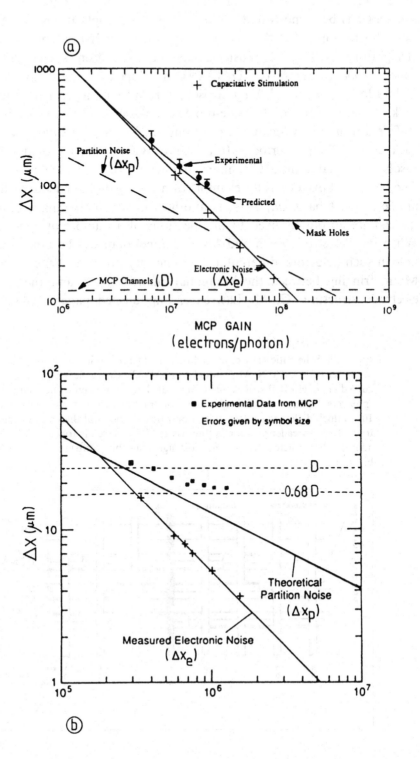

limitations of two-amplifier-per-axis charge division. To date, their development has been motivated by high count rate applications in UV and
optical astronomy, but their conversion to the soft X-ray band would be
straightforward. Fig. 3.22 illustrates the operation of one such system, the
coded anode converter (CODACON), developed at the University of
Colorado for planetary UV spectroscopy from sounding rockets (McClintock *et al.*, 1982). In this one-dimensional device N code tracks are
sufficient to uniquely locate the positions of 2^N charge spreaders. McClintock *et al.* (1982) describe a 10-bit CODACON which provides 25 μm
resolution, equal to the MCP channel diameter, over a 25 mm field-of-view.
The basic CODACON is an example of a purely digital encoder (the pin
anode array, Oba *et al.*, 1979, is another) in which the image elements
(pixels) are absolutely fixed to the locations of a number of conductors.
When inter-electrode pitch and MCP channel diameter become comparable in such detectors, differential non-linearity problems arise related to
Moiré fringing between the hexagonal channel array and the readout
electrode structure (G. M. Lawrence, unpublished paper, 1986). 'Flat

Fig. 3.22. Schematic view of an eight-channel (three code track) coded anode
converter (CODACON; McClintock *et al.*, 1982). Charge leaving a single
curved-channel MCP strikes one of a parallel array of conducting charge
spreaders. Signals are induced on the code tracks (sited on the opposite face of
10 μm thick dielectric substrate) in proportion to the local thickness of the code
track. Each spreader position is thus assigned a reverse binary (Gray code)
address, shown here in the order: most significant bit (MSB) to least significant
bit (LSB).

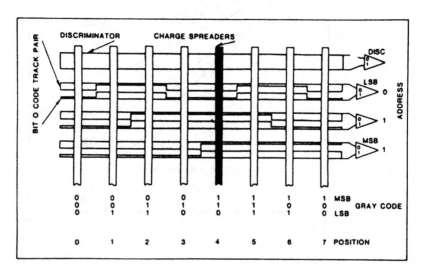

fielding' the detector response in the presence of the inevitable temperature induced mechanical shifts between MCP and readout element then becomes a serious problem.

Perhaps the most highly developed 'digital' readout system is the coincidence anode array encoder used in the *multi-anode microchannel array* (MAMA) detectors built by the Center for Space Sciences and Astrophysics at Stanford (Timothy, 1986; Timothy and Bybee, 1986). Coincident detection of the charge cloud emanating from a single curved channel MCP (fig. 3.23a) by interleaved coarse (C) and fine (F) position electrodes (fig. 3.23b) leads to the definition, in each dimension, of a large number (M^2) of pixels with a relatively small number ($2M$) of amplifier/discriminator chains. Here, M denotes the number of coarse or fine electrodes. Thus, for example, only 128 signal channels, plus the relevant high-speed address encoding logic, are required to create 1024 by 1024 pixel images. Pixel size in the present generation of MAMA detectors is $25 \times 25 \, \mu m^2$, a factor of two larger than the curved microchannel diameter in order to avoid Moiré patterns in the detector flat-field response. To deal with the spread of the MCP charge cloud in the $50 \, \mu m$ wide MCP–anode gap, MAMA detectors are 'partially centroiding'. That is, programming of the electronics allows selection of both two- and three-fold coincidence events (fig. 3.23b), so that positional information is obtained in fundamental 'half-pixel' bins in each axis. Fig. 3.23c and d illustrate, respectively, the fwhm resolution and positional sensitivity of a 256×1024 element MAMA array.

One final digital readout method which is currently making the transition from ground-based optical astronomy to space applications uses CCDs or silicon diode arrays (section 4.6). The concept of the 'intensified' silicon detector appears to have originated with Riegler and Moore (1973). In that work, as in subsequent research (Read *et al.*, 1986; Roberts *et al.*, 1986; Torr *et al.*, 1986), charge from a microchannel plate stack was allowed to fall on a phosphor-coated fibre optic coupling plate, which channelled light onto a position-sensitive silicon detector. Such MCP detectors which, with their multiple energy conversions, somewhat resemble the phosphor-based imagers of section 5.3.2., are designed, like the CODACON and MAMA tubes, for high resolution imaging at high count rates ($> 10^6$/s). Apparent limitations, such as the limited image areas ($\sim 1 \, cm^2$) and relatively large pixel sizes ($> 20 \, \mu m$) of commercial CCDs, can readily be overcome. The first, by splitting the image field between a number (up to 20 in a detector proposed for the Lyman mission, Roberts *et al.*, 1986) of coherent fibre optic bundles and the second, by spreading the phosphor light flash over many individual CCD pixels to allow the use of centroiding algorithms on the output pulse train.

The first satellite-borne intensified CCD detector was the 1200–1800 Å UV auroral imager built by the University of Calgary for the Swedish Viking mission (Anger *et al.*, 1987).

3.4.2   *The Einstein High Resolution Imager*

The scientific and technical characteristics of the Einstein Observatory HRI, developed at the Harvard-Smithsonian Center for Astrophysics, are summarised in Table 3.3. Two views of the HRI are shown in figs. 3.24a and b. The eventful development history of the HRI, including the role played by the Leicester X-ray Astronomy Group, is recounted by Tucker and Giacconi (1985). Although four sounding rocket flights of prototype HRIs ended in failure, the fifth succeeding only four months before the launch of HEAO-B, in-orbit operation of the instrument was relatively problem free. The 0.2–4 keV maps produced by the HRI remain the highest resolution images of the X-ray sky available at the time of writing.

Prelaunch, the active detector elements (UV/ion shield, MCPs and orthogonal wire grid readout; fig. 3.24b) were enclosed in a vacuum housing pumped by a small, integral ion pump (fig. 3.24a). In orbit, a motorised vacuum door opened to admit X-rays from the high-resolution telescope. Inflight calibration optics could be used to project UV light through a pattern of pinholes onto the MCP photocathode, so allowing checks to be made of the stability and resolution of the X-ray image readout. Each of the three flight HRIs (see Table 3.3) weighed 17 kg,

Fig. 3.23. Operation of a multi-anode microchannel array (MAMA) detector (Timothy, 1986; Timothy and Bybee, 1986). (a) Schematic view of MAMA detector with multilayer coincidence anode. *Key*: 1, curved-channel MCP (input, −2000 V; output, 0 V); 2, coincidence anode array; 3, quartz substrate; 4, upper plane coding electrodes (+75 V); 5, lower plane coding electrodes (+75 V); 6, SiO_2 insulating layer. Figures in brackets indicate operating potentials of the various components. (b) Allowed and rejected events in MAMA coincidence array. All events which stimulate four or more adjacent electrodes, or a lesser number of non-adjacent electrodes, are rejected. This facility could potentially be used to suppress noise due to extended particle tracks (section 3.3.6). (c) One-dimensional slices through MAMA image of a point source. One pixel corresponds to a linear distance of 25 μm. (d) Variation of measured centroid position with true spot position, illustrating positional sensitivity (section 3.3.5) at the sub-pixel level. (*Courtesy J. G. Timothy.*)

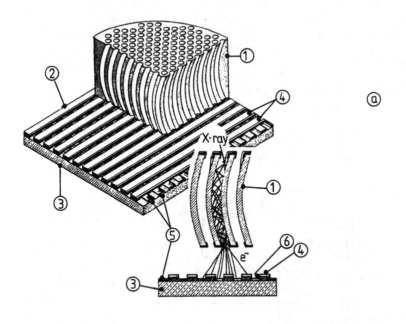

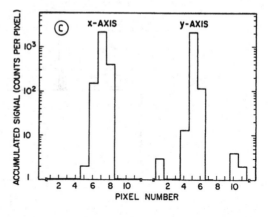

(a)

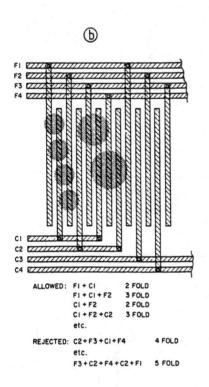

(b)

ALLOWED: F1 + C1   2 FOLD
          F1 + C1 + F2 3 FOLD
          C1 + F2   2 FOLD
          C1 + F2 + C2 3 FOLD
          etc.

REJECTED: C2 + F3 + C1 + F4   4 FOLD
           etc.
           F3 + C2 + F4 + C2 + F1   5 FOLD

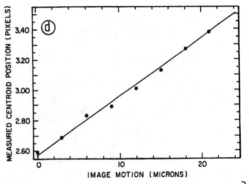

Fig. 3.24. Einstein observatory High Resolution Imager. (a) Outline drawing of the flight instrument (Henry *et al.*, 1977). (*Courtesy S. S. Murray.*) (b) Exploded view of detector components (Fraser, 1984, after Harrison and Kubierschky, 1979). UV/ion shield omitted for clarity. Charge collector bias voltages: 1, MCP output face earthed; 2, first grid $+250$ V; 3, second grid $+251$ V; 4, reflector plane $+200$ V. Grid plane separation $= 200\,\mu$m. Grid plane–reflector plane separation $= 300\,\mu$m.

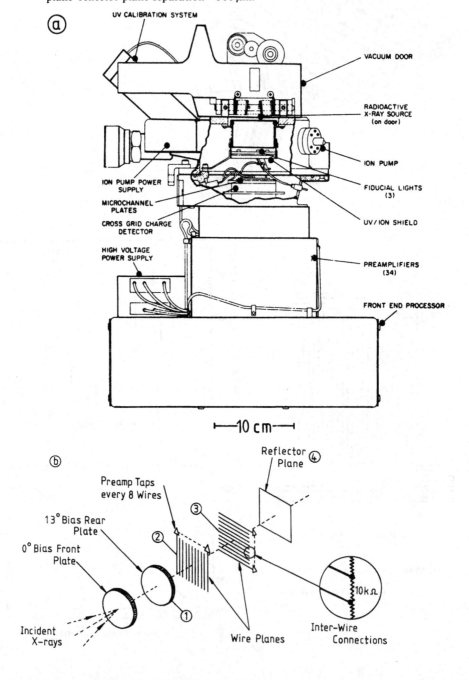

consumed 13.5 W in operation and fed signals into a central electronics assembly weighing 24 kg.

The key elements of the HRI were the *crossed grid charge detector* (CGCD), a pair of orthogonal wire grids terminated by a 16-fold partitioned *RC* line, and its associated electronics, described at the component level by Harrison and Kubierschky (1979). The wire planes consisted of 100 µm diameter wires on 200 µm centres, interconnected by 10 kΩ thick film resistors. Every eighth wire was connected to a preamplifier, necessitating 17 preamplifiers to cover the 25 mm active diameter of the detector. The presence of a metal reflector plane, biased less positively than either grid, ensured full collection of the MCP charge cloud. The small relative bias of the grids (fig. 3.24b) was chosen to produce 50:50 division of the charge cloud between image axes (Kellogg *et al.*, 1976). Fine position encoding was by the three-amplifier algorithm discussed in section 2.4.1.2. Subtle problems in the implementation of this algorithm (Harrison and Kubierschky, 1979) caused discontinuities (gaps) to appear in all raw HRI images, located at the positions of the preamplifier wires. Images such as that of the Cas A supernova remnant (fig. 1.13) featured, therefore, in their

Table 3.3. *Scientific and technical characteristics of the Einstein Observatory HRI (operational 1978–81)*

Kellogg *et al.* (1976), Henry *et al.* (1977), Kubierschky *et al.* (1978), Giacconi *et al.* (1979), Harrison and Kubierschky (1979).

Detector geometry	Chevron pair of Mullard $L/D = 80{:}1$, $D = 12.5$ µm MCPs. Front plate, $\theta_B = 0°$; rear plate, $\theta_B = 13°$. Interplate gap $d_G = 38$ µm.
Active area	25 mm diameter
UV/ion shield[a]	(1) 1.13 µm par N + 0.126 µm Al
	(2) 1.48 µm par N + 0.05 µm Al
	(3) 0.72 µm par N + 0.054 µm Al
Photocathode	4000 Å MgF$_2$, deposited at 4° to normal
Bandwidth (δE)	0.2–4 keV
Background rate (B_i)	0.2/cm^2 s (laboratory); <0.4/cm^2 s (in orbit)
Energy resolution ($\Delta E/E$)	No intrinsic energy resolution
Temporal resolution	7.8125 µs
Encoding method	Crossed grid charge divider: partitioned *RC* line with 17 amplifiers per axis
Spatial resolution (Δx)	33 µm fwhm ($= 2''$ in telescope focus)
Maximum count rate	Up to 500/s for undegraded imaging. In orbit, 100/s (telemetry limited)

[a] There were, in fact, three HRIs on the Einstein focal plane carousel, identical but for the detailed composition of their UV/ion shields.

raw state, a dark mesh of apparently missing data which had to be reconstructed ('degapped') before scientific analysis could begin.

Although the MgF$_2$-coated HRI had no intrinsic energy resolution, it could be used in conjunction with either movable bandpass filters or transmission gratings (section 1.4) to obtain spectral information. The so-called *broad-band filter spectrometer* (BBFS) consisted of Al/par N and Be/par N filters sited in the converging X-ray beam behind the grazing-incidence telescope. Either or both of these could be placed in the beam, to divide the spectrum into three bands: $E < 0.28$ keV, $0.8 < E < 1.5$ keV and $E > 1.5$ keV. Neither the BBFS nor the *objective grating spectrometer* (OGS) described in section 1.4 was in fact widely used during the Einstein mission, partly because of the ever-present fear of satellite mechanisms jamming in inconvenient positions (as indeed happened with one EXOSAT transmission grating) and partly because of their low sensitivity ($QA_s \sim 1$ cm^2 maximum for both devices). Fig. 3.25, however, illustrates the response of the HRI in spectral mode, for one of the few objects (the binary star Capella, α Aur) for which OGS data was obtained.

The success of the Einstein HRI has led to the incorporation of an almost-identical detector into the ROSAT main telescope instrumentation, as part of the US involvement in that German project. The ROSAT HRI, built under contract to NASA by the Harvard-Smithsonian group, will be used in the second, pointed, phase of the ROSAT mission. Apart from small changes in mechanical configuration, the only difference between HRIs is the substitution of CsI for MgF$_2$ (section 3.3.3) as the photo-cathode material for the ROSAT detector. This has resulted in an energy-dependent efficiency enhancement of between 1.5 and four times over the 0.1–2 keV band (Pfeffermann *et al.*, 1987). Ground test data (fig. 3.26) also shows that the spatial resolution of the ROSAT instrument ($\Delta x = 20$ μm) is (unaccountably) rather better than that of its predecessor.

A second direct descendant of the Einstein HRI is the high resolution camera (HRC) being developed for the AXAF mission (section 1.5) by the Harvard-Smithsonian researchers in collaboration with the Leicester X-ray Astronomy Group and the University of Hawaii (Murray and Chappell, 1985). The HRC, however, differs radically from the original HRI in several respects. Use of 10×10 cm^2 Mullard MCPs will increase the image area almost 20-fold, while extension of the CGCD encoder to 65 amplifiers per axis will preserve the spatial resolution at the HRI level. The use of thick CsI coatings and partially saturated MCP operation (section 3.3.4) will impart a measure of energy resolution to the HRC, while the possible use of low-noise glass in MCP fabrication (section 3.3.6) and the certain use

Fig. 3.25. Einstein Observatory spectrum of Capella, dispersed by the 1000 lines/mm transmission grating onto the surface of the HRI (Mewe *et al.*, 1982). (a) Raw spectrum in 0.375 Å bins. The central peak corresponds to the undispersed zeroth order. (b) Spectrum after subtraction of zeroth order and background contributions, and superposition of +1 and −1 orders. Solid line: best-fit two-temperature line and continuum spectrum (Mewe and Gronenschild, 1981). (*Courtesy R. Mewe.*)

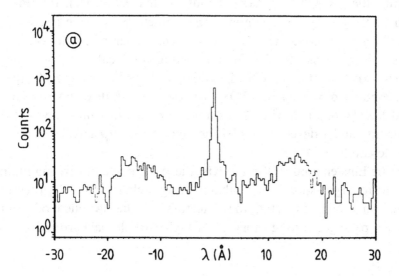

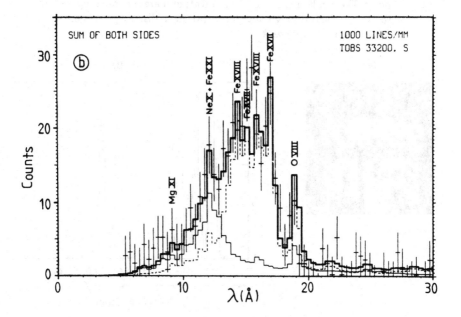

of plastic scintillator anti-coincidence shielding will substantially reduce detector background in orbit.

3.4.3   *Detectors for the ROSAT Wide Field Camera (WFC)*

The scientific and technical characteristics of the WFC detectors, for which the Leicester X-ray Astronomy Group has principal responsibility within the UK wide field camera consortium (section 1.5), are summarised in Table 3.4. Fig. 3.27a shows the complete WFC experiment, while fig. 3.27b is a sectional view of the detector assembly. The WFC was proposed in response to the German authorities' call, in 1980, for an ancillary experiment on ROSAT, and will complement the X-ray sky survey conducted by the main ROSAT telescope with its own survey in the 60–200 Å XUV band. A wide field sounding rocket instrument was at that time already under development by the Leicester group and the Center for Space Research at MIT.

Like the Einstein and ROSAT HRIs, the WFC detectors are maintained under vacuum (pressure $<10^{-5}$ mbar) during prelaunch tests by their own integral ion pump. In orbit, the detectors will be switched off during spacecraft passages through regions of high particle background such as

Fig. 3.26. ROSAT HRI test images (Pfeffermann *et al.*, 1987). (a) Image of test mask. The smallest slits are 12.7 μm wide and are separated by 50.8 μm.
(b) Image slices through fine slits of (a). Each display pixel is 6.8 μm wide.
(*Courtesy M. V. Zombeck.*)

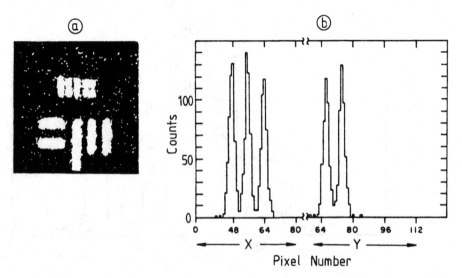

the South Atlantic Anomaly and the auroral zones. The shutdown signal will be provided by two background monitor detectors (fig. 3.27a) sensitive to low- and high-energy charged particles. During normal operation, electrons 'swept up' by the WFC telescope will be diverted from the active area of the detector by a concentric ring of permanent magnets situated between the XUV mirrors and the focal plane. As with the HRIs, a UV calibration system (Adams *et al.*, 1987) will be used to project a rectangular array of 100 μm diameter pinholes on to the MCP photocathode surface to check on the relative mechanical alignment and image calibration of the detectors. The complete WFC focal plane assembly, including the two identical detectors mounted on their turret exchange mechanism and a single eight-position filter wheel, weighs some 27.5 kg and consumes 3.7 W in steady state operation.

The response of the WFC detectors is greatly enhanced at the 30° angle of incidence of photons arriving from the optics by the use of both a CsI photocathode and an electron-collecting repeller grid (fig. 3.14d). The unique feature of these detectors, however, is their spherical curvature to match the focal surface of the telescope optics. Both the MCPs (fig. 3.5) and the resistive anode readout element are spherically curved to a common (165 mm) radius. Once the decision to match the detector to the

Table 3.4. *Scientific and technical characteristics of the ROSAT Wide Field Camera detectors*

Barstow *et al.* (1985), Wells (1985), Barstow and Pounds (1987).

Detector geometry	Chevron pair of $L/D = 120{:}1$, $D = 12.5\,\mu m$ Mullard MCPs. Radius of curvature $= 165\,mm$. Front plate, $\theta_B = 0°$; rear plate, $\theta_B = 13°$. Interplate gap $d_G = 160\,\mu m$
Active area	45 mm diameter
UV/ion shields	See Table 3.2
Photocathode	14 000 Å CsI, deposited at 11° to normal
Bandwidth (δE)	Determined by filters (Table 3.2) Survey mode $= 0.06$–$0.2\,keV$ in two bands
Background rate (B)	$0.6/cm^2\,s$ (laboratory); expected flight values in survey mode determined by XUV diffuse background (B_d) and geocoronal filter leakage ~ 3–$6/cm^2\,s$
Energy resolution ($\Delta E/E$)	No intrinsic energy resolution
Temporal resolution	5–40 ms
Encoding method	Curved resistive anode, amplitude ratio encoding
Spatial resolution (Δx)	100 μm fwhm ($= 0.66'$ in telescope focus)
Maximum count rate	200/s (normal mode); 400/s (fast mode)

Fig. 3.27. The ROSAT Wide Field Camera. (a) Schematic view. *Key*: 1, coaligned CCD star tracker; 2, experiment door; 3, nest of three coaxial, confocal Wolter–Schwarzschild Type 1 grazing-incidence mirrors, manufactured from aluminium and gold-coated for optimum reflectivity; 4, CEM and Geiger–Müller background monitor detectors, UV calibration system; 5, electron diverter; 6, filter wheel; 7, interchangeable MCP detectors on turret mechanism. (b) Detector cross-section. *Key*: 1, stainless steel outer body; 2, cover; 3, high-voltage feedthrough; 4, inner body; 5, curved resistive anode; 6, insulating spacer; 7, repeller grid; 8, curved MCP chevron. (*Courtesy A. A. Wells.*)

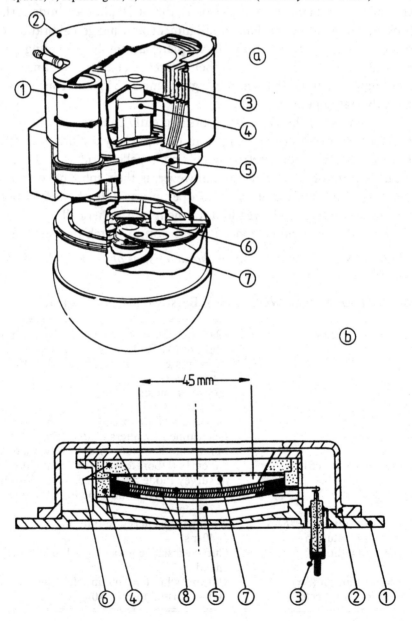

Fig. 3.28. WFC development detector imaging performance (Fraser *et al.*, 1988a). (a) C K X-ray image of 2×2 mm² area near detector centre. The linear structure corresponds to the shadow pattern cast by two randomly orientated 40 μm diameter wire grids preceding the MCP input. One of the grids acts as a repeller grid (section 3.3.3), the other as an earth plane. (b) Section through image (a) along the line *ab*. The valley-to-plateau count rate ratio indicates a fwhm detector resolution 2.5 times the wire diameter, i.e. 100 μm.

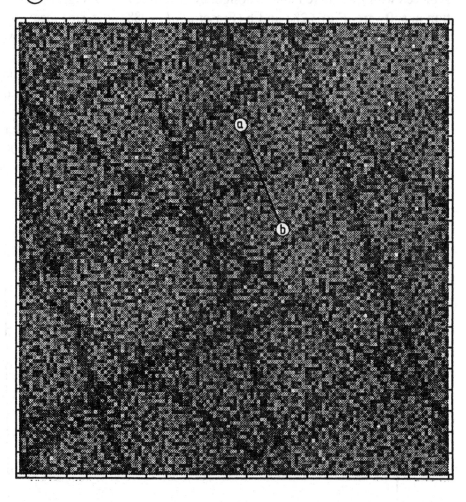

optics was taken, curving the readout element was considered an easier option than trying to accommodate the probable image distortions associated with combining curved MCPs and a planar encoder. The anode design is essentially that of fig. 2.20c, conformally mapped onto a spherical machinable ceramic former. The 55×55 mm² thick-film resistive layer itself is laid down by a screen printing process. The vertices of the 'square' image area lie some 4.6 mm above the centre of the anode. Barstow *et al.* (1985) have shown that relatively low integral distortion values ($\delta = 1\%$) can be achieved with this curved anode design, even prior to any image linearisation in software. Figs 3.28a and b (Fraser *et al.*, 1988a) describe measurement of the spatial resolution of the WFC detectors by the wire shadow technique (section 2.2.5).

During the second, pointed, phase of the ROSAT mission, an electronic 'zoom' mode will be used to restrict the detector field of view to the central 25 mm diameter, in effect halving the resolution contributions Δx_a and Δx_d.

On conservative assumptions regarding the opacity of the interstellar medium, the WFC all-sky survey is likely to detect upwards of 3500

Fig. 3.28. (Continued)

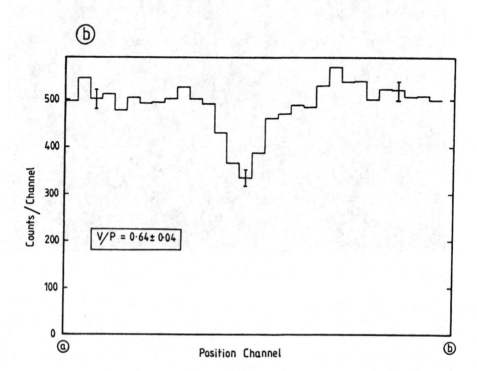

V/P = 0·64± 0·04

Counts / Channel

Position Channel

sources, mostly cool main sequence stars, white dwarfs and related objects (Barstow and Pounds, 1987), many of which will be previously uncatalogued objects.

3.5   Applications

Table 3.5 lists some of the many areas which have already benefited from research into microchannel plate detectors for X-ray astronomy, together with fields which may so benefit in future. The transfer of technology is in some instances rather direct – the use of high-efficiency

Table 3.5. *Some applications of microchannel plate detectors*

Application	Detected quantity	Reference
Soft X-ray microscopy	0.25–0.5 keV X-rays	Michette (1984)
Beam-foil spectroscopy	XUV emission from MeV ion collisions with thin foils	Demarest and Watson (1987)
Plasma diagnostics using crystal spectrometers	0.5–1 keV X-rays	Hailey et al. (1985); Duval et al. (1986)
Extended absorption fine structure (EXAFS)	0.2–0.8 keV X-rays	Gerritsen et al. (1986)
Plasma diagnostics using pinhole cameras	20–100 keV X-rays	Wang et al. (1985)
Medical imaging	10–150 keV X-rays	Gould et al. (1977)
Positron emission tomography	511 keV γ-rays	McKee et al. (1985)
Optical astronomy	Electrons from S20 photocathode	Kellogg et al. (1979); Rees et al. (1980); Firmani et al. (1982); Allington-Smith and Schwarz (1984); Clampin et al. (1987)
Low-energy electron diffraction (LEED)	100 eV electrons	Stair (1980)
Electron spectroscopy	Electrons	Van Hoof and Van der Wiel (1980); Richter and Ho (1986)
Proton beam diagnostics	Electrons from converter cathode	Cline et al. (1979)
Neutron radiography	Electrons from boron converter	Mullard Tech. Info. (1981)
Ion spectroscopy	3–25 keV positive ions	Hellsing et al. (1985)
Ion charge analysis	n+ noble gas ions	Hanaki et al. (1983)
Plasma analysis from the Giotto satellite	Electrons/positive ions	Balsiger et al. (1987); Johnstone et al. (1987); Reme et al. (1987); Wilken et al. (1987)
Detection of minimum ionising particles	MeV βs, GeV/c pions and protons	Bateman (1977b); Oba et al. (1981)

MCP coatings, for example, in ground-based X-ray detectors. Elsewhere, the channel plate geometries and readout methods intensively developed for X-ray and XUV astronomy have been adapted for the detection of electrons or positive ions or other particles. The electron and ion detection efficiencies of microchannel plates are discussed by Fraser (1983b) and by Gao *et al.* (1984), respectively.

4

Semiconductor detectors

4.1   Introduction

This chapter describes the use in X-ray astronomy of semiconductor ionisation detectors as non-dispersive spectrometers of high energy resolution. Semiconductor detectors which operate on a calorimeteric principle are described separately, in Chapter 6.

The early history of semiconductor radiation detectors is comprehensively described by McKenzie (1979). The first practical 'solid state' detectors – the term is usually taken to exclude scintillation counters (Chapter 5) – were small germanium surface barrier devices with gold electrodes (section 4.2.1), fabricated in the late 1950s. Such devices could be regarded as solid state analogues of the gas-filled ionisation chamber (Wilkinson, 1950), in that the primary ionisation produced in the dielectric and collected, without multiplication, by its internal electric field, was found to be proportional to the energy of an incident particle.

Over the past 30 years, improvements in material purity (Eichinger, 1987) and advances in microelectronic process technology have given rise to an array of detector types based on electron–hole pair creation in cooled silicon or germanium or in a number of 'room temperature' materials, of which the most developed is currently mercuric iodide. As in the case of gas-filled detectors (Chapter 2) much of the impetus for the new semiconductor detectors comes from the particle physics community (Kemmer and Lutz, 1987). Trends in modern particle physics include the development of both large (1 m²) silicon detector arrays (Borer et al., 1987) and of integrated detectors in which some at least of the signal processing is embodied 'on chip'.

In X-ray astronomy, however, solid state detector research is still driven primarily by the desire for high spectral resolution. Thus, for example, charge coupled devices (CCDs) and, increasingly, silicon drift chambers, both of which are imaging structures with low output capacitance, low readout noise and hence good energy resolution, are under active investigation as focal plane X-ray detectors, while silicon photodiode arrays (Fontaine *et al.*, 1987), with their much higher output capacitances, are not. Similarly, silicon avalanche photodiodes (Webb and McIntyre, 1976; Squillante *et al.*, 1986), the solid state analogues of the gas proportional counter, have never found application in X-ray astronomy because the presence of internal gain in any detector implies the presence of additional sources of energy blur, due to the statistical fluctuations in, and spatial non-uniformity of, the avalanche process.

In this chapter, the operating principles common to all semiconductor X-ray detectors are first outlined (section 4.2) before the devices of interest in astronomy are described in detail. Section 4.3 deals with soft X-ray silicon diodes, both monolithic and segmented, while sections 4.4 and 4.5 describe, respectively, the use of germanium and mercuric iodide detectors for harder fluxes. The final sections of the chapter are concerned with the imaging semiconductor detectors introduced above – CCDs (section 4.6) and silicon drift chambers (section 4.7). Each section includes descriptions of flight instruments, past and proposed.

4.2   Physical principles of semiconductor detector operation
4.2.1   *X-ray interactions and charge transport*
All solid state X-ray detectors consist of a volume of semiconducting material, subdivided by impurity doping into regions of differing conductivity, within which a charge collecting electric field can be established by the application of appropriate bias voltages to a set of surface contacts.

The usual result of soft X-ray absorption in a semiconductor is the creation of $N = E/w$ electron–hole pairs, analogous to electron–positive ion pairs in a counting gas (eq. 2.1). The K shell (binding energy 1.84 keV) fluorescence yield of silicon is only 5% (fig. 2.2). As in a gas, both negatively and positively charged charge carriers (a hole is simply an electron vacancy in the valence band of the solid) are free to move, in opposite directions, under the influence of an applied electric field. For both silicon and germanium (Table 4.1), the energy required to create an electron–hole pair, the ionisation energy w (eq. 2.1), is approximately 3 eV, so that roughly ten times as many charge carriers are liberated by the absorption in these media of a photon of energy E than in a counting gas. w is a weak function of temperature, varying in silicon from 3.62 eV at room temperature to 3.7 eV close to absolute zero (Canali *et al.*, 1972). The temperature variation of ionisation energy in germanium is described by Pehl *et al.* (1968).

Statistical fluctuations in N in semiconductors, as in gases, can be quantified in terms of a Fano factor F (eq. 2.2). Experimental Fano factors in silicon and germanium, determined by extrapolation from observed detector energy resolutions (Lumb and Holland, 1988a) and therefore to be regarded truly as upper limits, are nevertheless some two or three times less (Table 4.1) than the equivalent figure of merit in any counting gas. Thus, not only are more charge carriers generated in semiconductor detectors per kiloelectronvolt deposited energy than in gases, the variance in N is much less. Semiconductor detectors are therefore capable of much better energy resolution (section 4.2.4) than any form of gas counter.

Creation of free charge carriers by absorbed radiation represents a disturbance of the normal thermal equilibrium in the semiconductor. In a perfectly pure (intrinsic; suffix i) semiconductor the equilibrium density of electrons in the conduction band n_i is equal to the equilibrium density of holes in the valence band p_i and is given by the expression (Ewan, 1979):

$$n_i = AT^{3/2}\exp(-E_g/2kT)$$

where T is the absolute temperature, k is Boltzmann's constant and E_g is

Table 4.1. *Properties of semiconductor X-ray detection media*

Pehl and Goulding (1970), Canali *et al.* (1972), Ewan (1979), Knoll (1979), Haller (1982), Dabrowski *et al.* (1983), Di Cocco *et al.* (1985), Iwanczyk *et al.* (1985), Cuzin (1987).

Parameter	Si ($T=77$ K)	Ge ($T=77$ K)	HgI$_2$ ($T=300$ K)
Atomic number	14	32	80, 53
Density (g/cm³)	2.33	5.33	6.40
Melting point (K)	1683	1210	523
Dielectric constant κ	12	16	?
Band gap E_g (eV)	1.12	0.74	2.15
Electron–hole energy w (eV)	3.62	2.96	4.15
Fano factor F[a]	0.1	0.08	0.1
	(0.115)	(0.13)	(0.08)
Electron mobility–lifetime product $\mu_e\tau_e$ (cm²/V)	0.4	0.8	$\sim 10^{-4}$
Hole mobility–lifetime product $\mu_h\tau_h$ (cm²/V)	0.2	0.8	$\sim 10^{-5}$

[a] 'Consensus' experimental values, with theoretical predictions (Alig *et al.*, 1980) in parentheses.

the width of the forbidden energy gap between the (bound electron) valence band and the (free electron) conduction band. E_g has the characteristic value 1 eV in semiconductors (Table 4.1). A, finally, is a material constant, which includes the effective masses of electrons and holes (Haller, 1982). At room temperature, n_i is of order 10^{10} cm^{-3} and 10^{13} cm^{-3} for silicon and germanium, respectively.

The properties of even the purest available real semiconductors are dominated either by very small levels of residual metallic impurities or by deliberately introduced dopants. Substitution of a pentevalent atom, such as phosphorus, into tetravalent silicon or germanium produces 'donor' energy levels close to the top of the band gap and results in a conduction electron density n essentially equivalent to the donor impurity density. In such 'n-type' material free electrons are more plentiful than holes and are referred to as the majority carriers. If, on the other hand, a trivalent electron acceptor, such as boron or aluminium, is introduced, the energy levels created just above the valence band will all be filled, leaving holes as the majority carriers in a 'p-type' semiconductor. Very heavily doped layers are denoted n+ or p+. Compensated layers, containing equal amounts of donor and acceptor atoms, have properties very similar to those of pure intrinsic material and so are also denoted by the suffix i. In all cases, the product of electron and hole concentrations must equal its value in the intrinsic material. That is:

$$np = n_i p_i$$

The resistivity of a semiconductor, ρ_s, may be written in the form:

$$\rho_s = [e(n\mu_e + p\mu_h)]^{-1} \tag{4.1}$$

where e is the charge on the electron and μ_e, μ_h are, respectively, the electron and hole mobilities. By definition, the electron and hole drift velocities in a uniform field of strength E_d, are given by the expressions (cf. section 2.2.1):

$$W_e = \mu_e E_d; \ W_h = \mu_h E_d$$

W_e and W_h are of similar magnitude (10^6–10^7 cm/s) for most field strengths in silicon and germanium, while in a counting gas electron and positive ion mobilities may be radically different. As in a gas, charge carriers drifting in a semiconductor are subject to recombination and trapping (attachment) processes which may seriously degrade the spectroscopic performance of the detector. Impurities in the semiconductor may 'immobilise' holes or electrons for times longer than the readout time of the device, or may

sequentially capture both an electron and a hole, so prompting recombination. Both mechanisms contribute to finite lifetimes τ_e, τ_h for electrons and
holes in the material. The trapping lengths λ_e, λ_h, where

$$\lambda_e = \mu_e \tau_e E_d; \; \lambda_h = \mu_h \tau_h E_d \qquad (4.2)$$

then represent the average distances which photon induced charge carriers
may travel before being lost.

Equations (4.1) and (4.2) together impose important constraints on the
geometry of practical semiconductor detectors. The latter expressions set a
limit to the physical thickness of any detector. Obviously, the depth of a
charge collecting region can never be greater than the trapping length. For
silicon and germanium operated with fields in the range 100–1000 V/cm,
this presents no practical problem (Table 4.1). Even for mercuric iodide,
where detector thicknesses are limited to about a millimetre by trapping
and recombination effects, soft X-ray absorption efficiencies approaching
100% can still be achieved.

Suppose now one tries to construct an X-ray detector by simplemindedly
depositing conducting electrodes on opposite sides of a semiconducting
slab and applying a charge collecting field between them. Equation (4.1)
can be used to show that, in most circumstances, such an attempt is
doomed to failure. Even at the temperature of liquid nitrogen ($T = 77$ K),
the resistivities of most available silicon or germanium wafers (up to
~ 20 kΩ cm for Si: von Ammon and Herzer, 1984) are too low, and the
consequent steady state 'leakage currents' far too high, to permit the
detection of the minute photocurrent associated with the absorption of a
single X-ray photon. Most practical silicon or germanium detectors,
therefore, consist of reverse biased p–n or p–i–n junction diodes (fig. 4.1),
across which, in principle, no leakage current can flow. The electrostatics of
such diodes can easily be determined by solution of Poisson's equation
(Knoll, 1979). In practical detectors leakage currents at 77 K, arising from
surface channels rather than in the bulk, lie in the picoampère range
(Goulding and Landis, 1982).

p–n junction diodes can be manufactured in several ways:

(a) By exposing a homogenous crystal of (say) p-type material to
a high-temperature vapour of n-type impurity – the diffusion
method.

(b) By bombarding the crystal with monoenergetic impurity ions from
an accelerator – the ion implantation method.

(c) By evaporating a thin metal/metal oxide p-type contact, typically of gold, onto an n-type semiconductor – the surface barrier method (fig. 4.1a).

In each case, it is the region in the vicinity of the junction itself which acts as the active volume of the detector. Within this so-called depletion region, an electric field is set up which sweeps out photo-induced charge carriers towards the sides of the junction in which they are the majority carrier. The width of the active region – the depletion depth d – is given approximately by the expressions

$$d = (2\kappa\varepsilon_0 V_G / e N_{AD})^{\frac{1}{2}} = (2\kappa\varepsilon_0 \mu \rho_s V_G)^{\frac{1}{2}} \qquad (4.3)$$

where κ is the dielectric constant of the semiconductor in question (Table 4.1), ε_0 is the permittivity of free space, V_G is the potential difference applied across the junction, N_{AD} is the dopant concentration on the side of the junction with the lower impurity level, and μ is the associated mobility. The depletion depth in a semiconductor detector is analogous to the depth of the drift region in a proportional counter (fig. 2.1).

A reverse biased p–n junction has an associated capacitance per unit area

$$C = \kappa\varepsilon_0 / d = (\kappa\varepsilon_0 e N_{AD} / 2 V_G)^{\frac{1}{2}} \qquad (4.4)$$

which must be minimised in order to minimise the electronic component of the detector energy resolution (section 4.2.4).

Equations (4.3) and (4.4) dictate that diode detector performance – both in terms of quantum detection efficiency and energy resolution – is improved by increasing V_G. Each diode, however, will have a maximum tolerable operating voltage above which high-field electrical breakdown will occur. The technique of 'lithium drifting' (Pell, 1960) is used in practice to produce deep (of the order of several millimetres) detection volumes.

The lithium drift process uses the fact that lithium, in common with the other alkali metals, forms interstitial electron donors in silicon and germanium crystals. Field diffusion of lithium ions into the slightly p-type

Fig. 4.1. Semiconductor X-ray detector geometries and internal electric field configurations. The arrows at left indicate the direction of the electric field vector. Hole and electron motion are indicated by dashed lines. (a) Surface barrier p–n junction. (b) Lithium drifted n–i–p junction. (c) High purity germanium detector. Residual impurities are p–type. (d) Double-sided silicon drift chamber. In addition to the parabolic depletion field in the z direction, a constant transverse drift field is applied in the x direction, via the external potential dividers, in order to channel electrons to the single readout anode. (e) Front-illuminated charge coupled device (CCD). Depletion region, field-free region and substrate thicknesses are not to scale. The maximum field strength in the depletion region may be several kilovolts per millimetre.

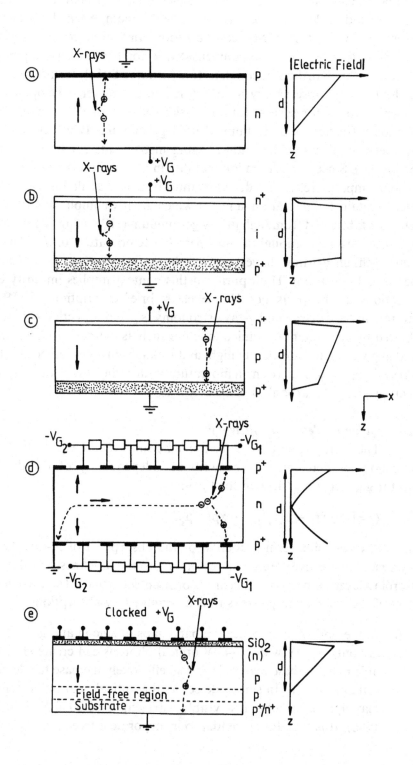

material which results from the best semiconductor processing produces, after a period of days or weeks (Knoll, 1979), a compensated region with extremely low nett impurity concentration and high resistivity. With contacts added (fig. 4.1b), this compensated region forms an n–i–p junction detector. Lithium drifted silicon detectors (denoted Si(Li)) formed the basis for the Einstein Observatory S³ detector, and are under development, in segmented form, for several future missions in 1–10 keV imaging X-ray astronomy (section 4.3). Lithium drifted germanium (Ge(Li)) detectors have been used in shielded, non-imaging instruments for harder X-rays (section 4.4). Since the lithium ion mobility in germanium is relatively high at room temperature, Ge(Li) detectors must be stored at all times at 77 K in order to retain compensation, a factor which greatly complicates their use. Detectors fabricated from high purity germanium (HP Ge: fig. 4.1c), whose resistances are high enough at an operating temperature of 77 K not to require lithium drifting, have been developed in order to overcome this operational constraint. High purity in this context implies impurity concentrations at the parts per 10^{12} level. A brief description of HP Ge detectors in hard X-ray and γ-ray astronomy is given in section 4.4, and is followed by an account of semiconductors such as mercuric iodide, whose band gap values are sufficiently high, and leakage currents sufficiently low, to permit operation at room temperature with either evaporated (palladium) or painted (carbon) metallic contacts.

4.2.2   *Quantum detection efficiency*

The X-ray quantum detection efficiency (Q in eq. 1.1) of any semiconductor detector may be written, independently of the detailed detector geometry (fig. 4.1), in the form:

$$Q=[\Pi_m \exp(-\mu_m t_m)][1-\exp(-\mu_s d)] \tag{4.5}$$

The first, composite, term accounts for the absorption of X-rays before they reach the active depletion region. t_m is the thickness of the mth absorbing layer, and μ_m is its linear absorption coefficient. The low-energy cutoff of the detector response is thus determined by absorption:

(i) In the optical filter. Since the band gaps of silicon and germanium are only ~ 1 eV wide, 1–4 eV optical photons can create electron–hole pairs in these materials and so effectively increase the detector leakage current. In silicon detectors the optical quantum efficiency may approach 80% at a visible wavelength of 5000 Å (Mackay, 1986). It is therefore essential to incorporate a free-standing metal or plastic UV/optical/near IR filter into silicon- or germanium-based detector designs, while the larger band gap of mercuric iodide (Table 4.1) alleviates, but does not eliminate, the problem of UV sensitivity.

(ii) In the electrode structure used to apply bias voltages to the detector. This is an important consideration in CCD detectors, where several overlapping electrode and insulating layers may be necessary to apply clocked pulses to adjacent pixels (section 4.6.1).

(iii) In any dead layer (figs. 4.1b, c, e) between the detector surface and the upper boundary of the depletion region. Whereas the thicknesses of optical filters and electrodes are controlled parameters during detector manufacture, the thickness of the dead layer, in which incomplete charge collection occurs, varies unpredictably from detector to detector. Techniques for dead layer thickness measurement utilise the discontinuity in absorption coefficient at the K edge of the detector crystal (Jaklevic and Goulding, 1971) and so-called X-ray doublet methods (Musket, 1974; Maor and Rosner, 1978).

The second term of eq. (4.5) is the absorption efficiency of the depletion region, depth d. μ_s is the linear absorption coefficient of the detector medium. Fig. 4.2 compares the mean absorption depths $1/\mu_s$ in silicon, germanium and mercuric iodide. We see that 3 mm of silicon (a typical depletion depth for commercial Si(Li) detectors: Table 4.2) provides 95% X-ray absorption efficiency up to about 25 keV. The same thickness of higher-Z germanium extends the energy band to nearer 60 keV, while mercuric iodide has even more stopping power above the I K absorption edge (33.17 keV).

Equation (4.5) determines what might be termed the 'spectroscopic' quantum efficiency of a semiconductor detector, which may be a lower limit to the measured counting efficiency. X-rays absorbed in 'field-free' regions below the depletion layer, in cases where it does not extend throughout the complete detector (fig. 4.1e), may give rise to 'partial' events in which not all the free charge is collected (section 4.6.2.1).

4.2.3 *Energy resolution*

The energy resolution of any semiconductor detector may be expressed as the quadratic sum of three independent terms:

$$\Delta E = 2.36\{\sigma_N^2 + \sigma_R^2 + \sigma_A^2\}^{\frac{1}{2}} \tag{4.6}$$

Fig. 4.2. Mean absorption depth $1/\mu_s$ as a function of X-ray energy in silicon, germanium and mercuric iodide. The vertical lines indicate: the K absorption edges of Si, Ge, I and Hg at energies of 1.84, 11.1, 33.17 and 83.1 keV, respectively; the L edges of Ge at 1.217, 1.248 and 1.414 keV; the L edges of I at 4.557, 4.852 and 5.188 keV; the M edges of Hg at 2.295, 2.385, 2.847, 3.28 and 3.562 keV; the MI edge of I at 1.072 keV. The departure from linearity in the case of Si above 50 keV is due to the onset of Compton scattering.

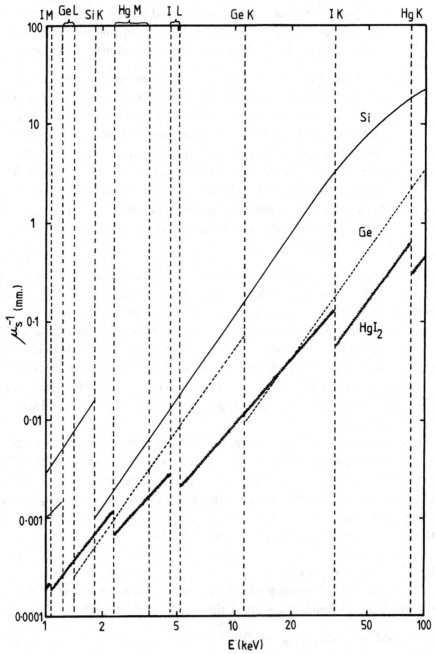

σ_N is the variance of N, the number of primary electron–hole pairs (section 4.2.1); σ_R arises from the loss of charge during collection, drift or transfer; and σ_A is the energy blur due to noise in the detector pre-amplifier, main amplifier and signal processing electronics. Both readout and amplifier terms are commonly expressed in terms of noise charges, denoted here by R, A, respectively, and measured in signal equivalent electrons. Equation (4.6) can therefore be rewritten in the form:

$$\Delta E = 2.36w\{FE/w + R^2 + A^2\}^{\frac{1}{2}} \qquad (4.7)$$

Obviously, the partial quadratic sum of R and A must be minimised if (i) the fundamental Fano factor limit to resolution is to be approached in practice, and (ii) the device is to function at all as a detector of sub-kilovolt X-rays.

Expressions for σ_A in terms of basic preamplifier parameters are given, for example, by Dabrowski (1982) and Di Cocco *et al.* (1985). σ_A is always linearly dependent on the input capacitance of the preamplifier, hence on the area of the detector. σ_A also depends on the residual leakage current of the detector and the temperature dependent characteristics of the *field effect transistor* (FET) invariably used as the first charge amplification stage. Semiconductor detectors of all kinds for use in soft X-ray astronomy, where energy resolution is the imperative factor, are, therefore, typically rather small ($< 1\,\mathrm{cm}^2$) devices, cooled to liquid nitrogen temperatures, with their amplifiers running somewhat warmer (110–150 K). In the case of CCDs with 'on-chip' amplifiers, differential cooking is not possible and the optimum operating temperature ($\sim 180\,\mathrm{K}$ for most devices) is partly determined by a tradeoff between leakage current and FET noise terms (section 4.6).

While the first and last terms in eqs. (4.6) and (4.7) describe symmetric

Table 4.2. *Some semiconductor detector manufacturers*

EG&G Ortec
100 Midland Road, Oak Ridge, Tennessee 37830, USA

Princeton Gamma-Tech
1200 State Road, Princeton, New Jersey 08540, USA

Canberra Industries Inc.
45 Grove Avenue, Meriden, Connecticut 06450, USA

Kevex Corporation
Foster City, California 94404, USA

Hamamatsu Photonics KK, Solid State Division
1126 Ichino-cho, Hamamatsu City 435, Japan

broadening of detected spectral lines, charge loss, by whatever means, is always an asymmetric blurring process which skews pulse height spectra toward low charge levels. Absorption in field-free regions of the detector, as described in the previous section, is one source of partial events. In the case of highly absorbing (fig. 4.2) germanium and mercuric iodide, a window effect (Dabrowski, 1982) – analogous to that observed in gas detectors (section 2.2.4) – can give rise to significant 'partial event plateaux' and indeed renders germanium unacceptable as a low-energy X-ray detection medium (Llacer *et al.*, 1977), despite the fact that its Fano factor limited resolution (fig. 4.3) is better than that of silicon.

Fig. 4.3. Energy resolution in semiconductor detectors. All measurements and predictions refer to operation at liquid nitrogen temperature unless otherwise stated. Measurements displayed at zero energy represent electronic component of resolution. *Key*: Open circles = 4 mm diameter Si(Li) detector (Lochner and Boldt, 1986); filled circles = 3 mm diameter Si(Li) detector (Musket, 1974); crosses = EEV P8603 CCD, isolated event data (Lumb and Holland, 1988b); filled squares = HgI_2 crystal at 273 K, amplifier at 77 K (Dabrowski *et al.*, 1983); open squares = Iwanczyk *et al.* (1985), 1 mm² HgI_2 detector, same arrangement; full curve = predicted response of silicon drift chamber at 100 K (Sumner *et al.*, 1988); broken curves = (experimental) Fano factor limit for silicon, germanium and mercuric iodide.

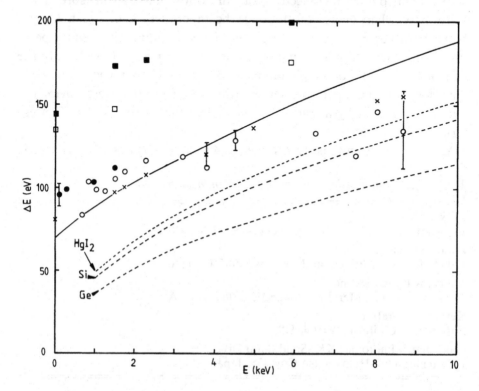

One final factor which may contribute to the energy resolution of a semiconductor detector is inhomogeneity of the crystal material (Dabrowski, 1982). Local variation in trapping length arising from material inhomogeneity can give rise to pulse height variation at the collecting electrode in detectors with a parallel-plate geometry (figs. 4.1a, b, c).

Fig. 4.3 summarises the 'state of the art' spectral resolution in various forms of solid state detector.

4.2.4   Dark noise

Provided N, the number of X-ray induced electron–hole pairs, is much greater than the equivalent electronic noise charge A, the count rate B_i (eq. 1.1) in a semiconductor detector is due to the prompt and delayed effects of the in-orbit charged particle and γ-ray background fluxes (section 2.3.2).

Unlike microchannel plates (section 3.3.6) and alkali halide scintillation counters, inbuilt internal radioactivity is not a problem with highly purified semiconductor crystals. Indeed, silicon detectors are commonly used to measure ultra-low α, β activity levels (Wojcik and Grotowski, 1980). In the case of CCDs, chip packaging has been suspected of contributing to ground-based background count rates (Thorne *et al.*, 1986), while glass windows in the low-temperature dewars housing CCDs for optical astronomy have been definitely found to be a source of unwanted background events (Mackay, 1986).

Of the three types of background rejection technique introduced in the context of large-area proportional counters in section 2.3.2, energy selection is the most commonly used in conjunction with soft X-ray semiconductor detectors. Energy deposition by charged particles in silicon has been very extensively studied because of the interest in silicon detectors for high-energy particle physics. Although the details of the collision processes are complex (Bichsel, 1985), to first order one may assume that minimum ionising particles create ~ 80 secondary electrons per micron of silicon track (Lumb and Holland, 1988b). Single fast particles would therefore deposit around 0.88 MeV in a 3 mm deep Si(Li) detector, far above the X-ray energy band of interest when such detectors are used in astronomy. Only with standard CCDs, where the depletion depth d may be less than 10 μm, do many particles deposit energies in the soft X-ray range (section 4.6.2.1).

Of the two remaining rejection techniques, anti-coincidence between elements of a subdivided detector has been demonstrated with Si(Li) detectors (section 4.3), HP Ge detectors (section 4.4) and CCDs (section 4.6.2.1) and proposed for mercuric iodide soft X-ray detector arrays (section 4.5). Anti-coincidence shielding by layers of scintillator is commonly used at higher energies (sections 4.4, 4.5). Rise-time discrimination is impracticable in semiconductor detectors because charge collection times d/W_e, d/W_h are so short (only of order 10^{-8} s even in deeply depleted Si(Li) or germanium detectors).

As pointed out in section 2.3.3, irradiation by charged particles inevitably leads to detector damage. The physics of radiation damage in semiconductor detectors is described by Kraner (1982) and by Van Lint (1987). More detailed discussions of in-orbit background and damage are given below for each individual detector type.

4.3   Si(Li) diodes

The first use of small (1 cm²) collimated Si(Li) diodes in cosmic X-ray astronomy was in an unsuccessful sounding rocket flight to observe Sco X-1 (Singer *et al.*, 1972). Two out of three liquid nitrogen cooled diodes were subjected during the flight to microphonic preamplifier noise arising from intermittent activation of the rocket's attitude control system. The door covering the third diode failed to open.

Research at the French Centre d'Etudes Nucleaires, Saclay, in the early 1970s (Griffiths *et al.*, 1975) culminated in the first (and, so far, only) successful sounding rocket flights of Si(Li) detectors (Schnopper *et al.*, 1982; Rocchia *et al.*, 1984). Two collaborative observations were made in 1980 and 1981 of the North Polar Spur, an old supernova remnant occupying a large fraction of the northern galactic hemisphere, which had previously been studied with thin-windowed proportional counters and GSPCs (Inoue *et al.*, 1979). Fig. 4.4 compares the quantum detection efficiency and energy resolution of a Si(Li) detector with those of a thin-window GSPC.

The solid state telescope flown by the Saclay and Harvard-Smithsonian groups consisted of three 1 cm diameter Si(Li) diodes mounted in a liquid nitrogen cryostat. A magnetic diverter (section 3.4.3) was used to reduce the particle background reaching the detectors through a coarse (1 or 0.25 sr) collimating diaphragm to the level 0.1 counts/cm² s keV (Briel *et al.*, 1988). Fig. 4.5 compares the Si(Li) spectrum of the North Polar Spur obtained during the first rocket flight with that obtained by an earlier GSPC experiment (Inoue *et al.*, 1979). The similar shapes of these spectra reflect the similarity in the energy resolution of the two detector types in the 0.1–1 keV band (fig. 4.4).

Successful though the Saclay rocket flights were, it can be shown from eq. (1.1) that the usefulness of any collimated Si(Li) detector, with its active area severely constrained by energy resolution requirements (section 4.2.3), is limited to the mapping of large-scale features in the diffuse X-ray background and perhaps the examination of the very few brightest point sources. The full power of a non-dispersive Si(Li) spectrometer is only realised when it is placed at the focus of an X-ray telescope. Here too, flight experience to date is limited to one instrument: the Einstein Observatory Solid State Spectrometer (S³), built by researchers at the NASA Goddard Space Flight Center, whose scientific and technical characteristics are summarised in Table 4.3.

The S³ detectors and their cryostat are shown schematically in figs. 4.6a and b. Each of the redundant detector packages consisted of a 9 mm diameter Si(Li) X-ray detector and a parallel high purity germanium diode

Fig. 4.4. Si(Li) detector quantum efficiency (left hand scale) and energy resolution (right hand scale) compared to those of a GSPC with a polypropylene entrance window (Rocchia *et al.*, 1984). Discontinuities in the Si(Li) quantum efficiency curve are due to X-ray absorption in an aluminium–parylene filter, gold bias electrode and silicon dead layer (section 4.2.2). Energies of astrophysically important emission lines are indicated. (*Courtesy R. Rocchia.*)

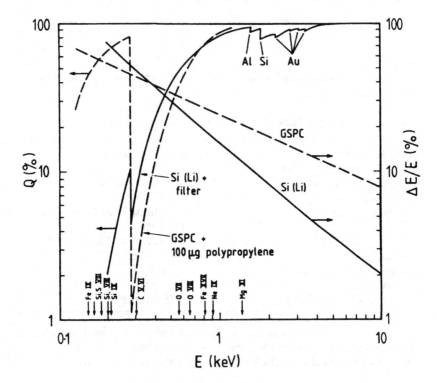

working in coincidence with it in order to reduce the particle background. The Si(Li) detectors were operated with the opposite polarity to that shown

Fig. 4.5. M-band (0.4–1.2 keV) X-ray spectra of the North Polar Spur. (a) Si(Li) spectrum, showing two-temperature model fit (Rocchia *et al.*, 1984). (*Courtesy R. Rocchia.*) (b) Thin-window GSPC spectrum (Inoue *et al.*, 1979). (*Courtesy H. Inoue.*) The common peak at 0.6 keV is due to a blend of O, C, N, Ne and Fe emission lines.

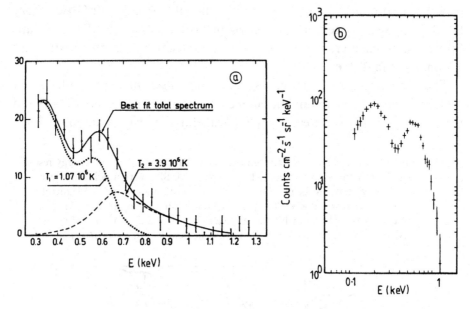

Table 4.3. *Scientific and technical characteristics of the Einstein Observatory S³*

Joyce *et al.* (1978), Giacconi *et al.* (1979), Holt *et al.* (1979).[a]

Si(Li) active area	6 mm diameter
Si(Li) depletion depth (d)	3 mm
Si(Li) bias voltage (V_G)	-600 to -1200 V
Si(Li) operating temperature (T)	100 K
UV/optical filter	Aluminised par N on 97% transparent Ni mesh
Effective area (QA_s)	~ 150 cm² at 1 keV
Bandwidth (δE)	0.5–4.5 keV[b]
Energy resolution ($\Delta E/E$)	$16/E\%$
Time resolution	2 µs–5 ms
Internal background (B_i)	0.28 counts/s

[a] There were in fact two identical Si(Li) detectors on the Einstein focal plane carousel, contained within the same CH_4/NH_3 cryostat (fig. 4.6b).
[b] Lower limit raised to 0.8 keV in orbit by accumulation of ice.

in fig. 4.1b, i.e. with the p side of the crystal – through which X-rays entered – at a high negative potential and the n side earthed. They were also mounted slightly out of the telescope focus in order to average out the X-ray transmission through their nominal 200 Å gold contacts, which were

Fig. 4.6. Einstein Observatory Solid State Spectrometer (S³: Joyce *et al.*, 1978). (a) Cross-section of detector housing, showing parallel Si(Li) and HP Ge detectors. X-rays enter from below. (© *1978 IEEE.*) (b) Cross-section of CH₄/NH₃ cryostat, with detector mount removed to the left for clarity. X-rays enter from the left. (*Courtesy S. S. Holt.*) (© *1978 IEEE.*)

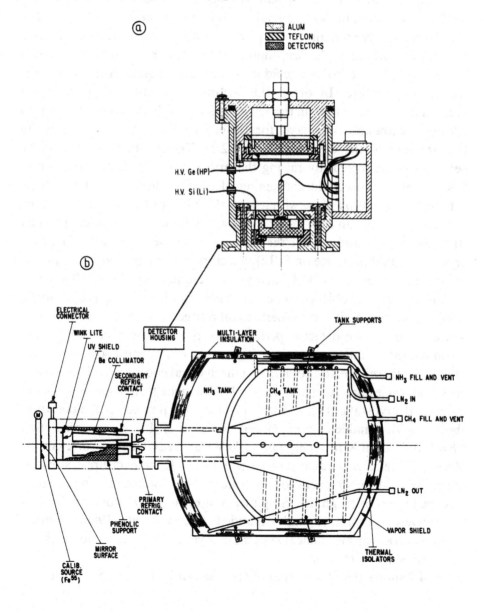

found to actually vary in thickness on a lateral scale of microns. The resulting variation of quantum efficiency with X-ray energy was very similar to that shown in fig. 4.4. The in-flight energy resolution ΔE, measured using 5.9 keV Mn K X-rays from an ^{55}Fe calibration source mounted on the instrument door (fig. 4.6b), was 160 eV – considerably higher than the best laboratory values for small Si(Li) detectors shown in fig. 4.3.

The S^3 operating temperature of 100 K was realised in orbit by means of a two-stage passive cooling system based on solid methane as the primary coolant and solid ammonia providing a secondary vapour shield at 150 K. The detectors were mounted on a primary cold plate directly connected to the central methane tank. Beryllium collimators connected to the secondary ammonia tank acted as a 'cold trap' for ambient gases evolved from the spacecraft structure. In orbit, this arrangement failed to prevent the accumulation of water ice on the active surfaces of the Si(Li) detectors, thus adding an extra absorbing layer for low-energy X-rays to those inherent in the structure of the device (section 4.2.2). The S^3 therefore had to be periodically 'defrosted' by raising the primary cold plate temperature to 220 K, with a consequent shortening of cryostat life. Even then, observations typically took place with a 1 μm thick layer of ice on the detectors, sufficient to mimic an interstellar hydrogen column density (see Appendix B) $N_H \sim 10^{21}/cm^2$ and cut off the detector response at 0.8 keV. The time dependent build up of ice on Si(Li) detectors used in ground-based *proton induced X-ray emission* (PIXE) analysis is described by Cohen (1987).

Condensation problems, together with problems of cryostat mass, volume and complexity, are generic to all refrigerated solid state detectors. Research into basic detector performance is therefore necessarily being accompanied by intensive engineering studies of in-orbit cooling systems. Multistage thermoelectric coolers and mechanical coolers based on Stirling or Solvay cycles are currently under development to replace or augment traditional cryostats. The former are currently limited to temperatures above 180 K. Madden *et al.* (1986) report a Si(Li) energy resolution $\Delta E = 190$ eV at an X-ray energy of 5.9 keV using a 200 K four-stage cooler. Mechanical refrigerators can easily reach the 100 K temperature desired for silicon-based detector operation, but inevitably produce vibration, which may result in microphonic amplifier noise, as their working fluid is alternatively compressed and expanded (Stone *et al.*, 1986). Neither thermoelectric nor mechanical systems can yet be considered qualified for long-term satellite missions.

Fig. 4.7 shows the S^3 spectrum of Capella (Holt *et al.*, 1979) which may be usefully compared with fig. 3.25, the Einstein Objective Grating Spectrometer (OGS) spectrum of the same star recorded in an observation more than four times as long. While the energy resolution of the solid state detector is insufficient to resolve any of the sub-kiloelectronvolt lines seen in the OGS data, its much greater bandwidth and higher quantum efficiency are also evident from comparison of the two figures.

The Einstein Solid State Spectrometer, like the Einstein HRI (section 3.4.2), has spawned a number of descendants which improve on the original in terms of focal plane coverage, position sensitivity and background rejection. The first of these to fly in space is likely to be Goddard Space Flight Center's Broad Band X-ray Telescope (BBXRT), part of the second Shuttle High Energy Astrophysics Laboratory (SHEAL-2) (Serlemitsos *et al.*, 1984). BBXRT consists of a quadrant nest of thin conical

Fig. 4.7. S³ spectrum of Capella (Holt *et al.*, 1979). The solid line represents the fit of a two-temperature model of an inhomogenous stellar corona to the data. Observation time = 7000 s. (*Courtesy S. S. Holt.*)

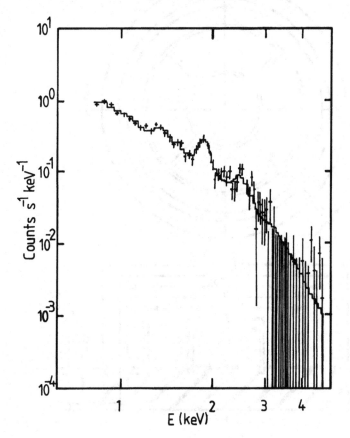

mirrors (Petre and Serlemitsos, 1985) with a segmented Si(Li) diode (fig. 4.8) at its focus. Mechanical or chemical grooving of a single crystal surface can be used to create close packed mosaics of independent detectors, each with its own amplifier and each capable of the energy resolution of an isolated diode of the same size ($\Delta E = 185\,\text{eV}$; $E = 5.9\,\text{keV}$). In the BBXRT design the central detector is intended for point source studies while the outer segments will be used (i) to provide simultaneous monitoring of the background during point source observations, (ii) to reduce the central detector background by pixel-to-pixel coincidence, and (iii) to generate coarse spectral maps of extended sources such as clusters of galaxies. Background reduction by the coincidence technique, although relatively

Fig. 4.8. Five segment Si(Li) detector. Top: plan view; bottom: section across a diameter. Dimensions in millimetres. (Serlemitsos *et al.*, 1984.) (© *1984 IEEE.*)

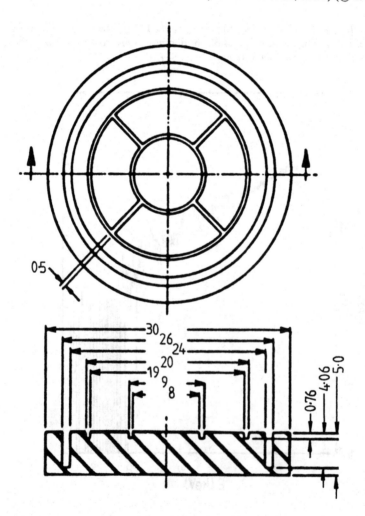

inefficient, is expected to reduce the BBXRT background to a level 5×10^{-4} counts/cm^2 s keV when augmented by a CsI shield around the cryostat (Serlemitsos *et al.*, 1984). Calibration techniques for the BBXRT detectors are described by Lochner and Boldt (1986) (see fig. 4.9).

A version of the BBXRT segmented diode is included in the Goddard/Wisconsin X-ray Spectroscopy Experiment on AXAF (Holt *et al.*, 1985; see section 6.3). A seven-fold Si(Li) detector has also been considered for the ESA XMM satellite payload (Briel *et al.*, 1988). For these long-term (ten year) observatory class missions, the stability of all solid state detectors under particle bombardment becomes an important issue. Si(Li) diodes are thought to be relatively 'radiation hard'. In particular, enhanced levels of charge trapping (section 4.2.1) do not lead to seriously degraded energy resolution because charge collection volumes are relatively small. Tolerable particle fluxes lie in the 10^{10}/cm^2 (5–50 MeV α-particles) to 10^{14}/cm^2 (2–5 MeV electrons) range (Kraner, 1982). Tolerable absorbed doses for typical detectors then lie between 1 and 100 krad (10^{-5}–10^{-3} J/g).

Fig. 4.9. Pulse height channel-to-energy calibration of a Si(Li) diode, illustrating the characteristic linearity of all solid state detectors (Lochner and Boldt, 1986). (*Courtesy J. C. Lochner.*)

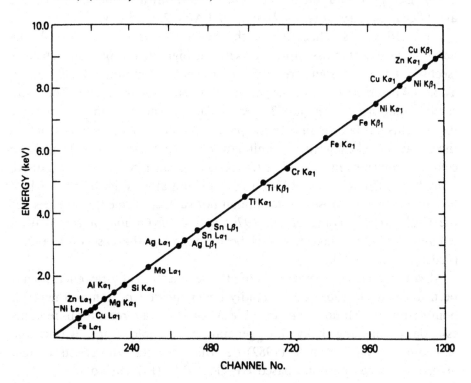

4.4   Ge(Li) and HP Ge detectors

Lithium drifted and, later, high purity germanium detectors have been extensively used in astronomy, as in the laboratory (Knoll, 1979), for high-resolution hard X-ray and γ-ray spectrometry. Non-imaging colli-mated germanium detectors have been used to perform detailed follow-up spectroscopy of line features discovered using alkali halide scintillation counters of much larger area but much poorer energy resolution (see section 5.2.2.1). Like scintillation counters, germanium detectors of either a planar (fig. 4.1) or a coaxial cylindrical geometry are essentially high-energy devices, with most of their bandpass beyond the scope of this monograph. In addition to the window effect exhibited at kiloelectronvolt energies (section 4.2.3), it is difficult to fabricate the large volume ($\sim 100\,\text{cm}^3$) germanium detectors desired for γ-ray studies with the thin surface dead layers necessary for good soft X-ray efficiency. Typical dead layer thicknesses in both HP Ge and Ge(Li) detectors lie in the range 0.3–0.6 mm (Peterson, 1975; Mahoney *et al.*, 1980). Typical low-energy detec-tion limits therefore lie in the range 20–50 keV (fig. 4.2). Upper limits to the bandpass, set by the depletion depth d, lie between extremes of 150 keV ($d=7\,\text{mm}$; Proctor *et al.*, 1982) and 10 MeV ($d=45\,\text{mm}$; Mahoney *et al.*, 1980). Energy resolution in such detectors is always dominated by the amplifier noise term A (eq. 4.7) at a level $\Delta E = 2.0\text{–}3.5\,\text{keV}$.

Fig. 4.10 shows schematically the key elements of a balloon-borne germanium detector assembly. As with all high-energy photon detectors (sections 4.5, 5.2) designs are driven by the need to minimise background. Table 4.4 describes some of the germanium detectors flown on balloon and satellite platforms in the last 20 years. The astronomical impact of these instruments has been mainly in the precise detection of γ-ray lines (such as the variable 511 keV positron annihilation line emanating from the galactic centre region (Leventhal *et al.*, 1978; Reigler *et al.*, 1985)) and of impulsive γ-ray bursts (Imhof *et al.*, 1974). Line emission at energies below 100 keV has, however, been observed (although not fully confirmed) in the case of the Crab Nebula (Ling *et al.*, 1979) and confirmation of scintillation counter X-ray line detection has been sought in the case of Her X-1 (Proctor *et al.*, 1982).

Although the energy resolution of HP Ge detectors at γ-ray energies has been found to degrade rather rapidly under in-orbit proton and neutron bombardment (Mahoney *et al.*, 1981), their X-ray resolution remains essentially unchanged by the generation of hole traps within the germanium crystal. Kraner (1982) describes in detail the effects of fast neutron damage in ground-based HP Ge and Ge(Li) detectors.

As with scintillation counters (section 5.2.2.1) proton induced radio-activity is an important source of background events in germanium detectors for astronomy. Dyer *et al.* (1980) have published a table of production rates for the major radioactive isotopes engendered by the particle bombardment of germanium. Coaxial HP germanium detectors have recently been produced, in which internal radioactive decays can be rejected by using coincidence between detector segments (Gehrels *et al.*, 1984; Varnell *et al.*, 1984). It is hoped that the total background count rate in future satellite detectors might thus be reduced ten-fold with respect to the levels observed during the HEAO-C mission (Mahoney *et al.*, 1980, 1981).

Fig. 4.10. Schematic view of balloon-borne germanium (Ge(Li) or HP Ge) detector assembly (see, for example, Leventhal *et al.*, 1977; Proctor *et al.*, 1982; Paciesas *et al.*, 1983). The collimator, side and lower anti-coincidence shields (ACS) are fabricated from either NaI(Tl) or CsI(Na) blocks. Collimator transmission $\theta_{\frac{1}{2}}$ is typically 5–10° fwhm at low energies. For a full description of the design of alkali halide background shields, see section 5.2.2.1.

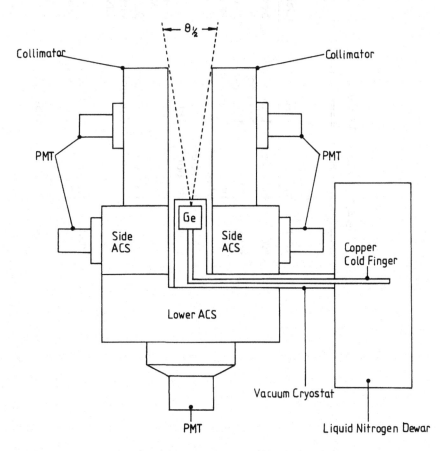

Table 4.4. *Lithium drifted and high purity germanium spectrometers.*

Year	Satellite/ Balloon	Institution	Detector	Bandpass (δE)	$\Delta E/E$ at 60 keV	Observation
1967	Balloon[a]	Univ. Calif., San Diego	5 cm³ Ge(Li)	—	—	Crab Nebula [1]
1972	USAF 1972-076B Satellite[b]	Lockheed Palo Alto	50 cm³ Ge(Li) in 130 K CO_2 cryostat	40 keV–2.8 MeV	~6%	γ-ray bursts, Sky survey [2,3]
1974	Balloon	JPL	4 × 50 cm³ Ge(Li) in LN_2 dewar	50 keV–10 MeV	~4%	Crab Nebula, diffuse emission [4,5]
1977	Balloon	Bell Labs., Sandia Labs.	92/130 cm³ Ge(Li) in LN_2 dewar	100 keV–5 MeV	—	511 keV emission from gal. centre [6,7]
1978	USAF P78-1 Satellite[c]	Lockheed, Palo Alto	2 × 85 cm³ HP Ge	50 keV–2.5 MeV	—	[8]
1978–81	Balloon	GSFC/Saclay/ Rice Univ.	3 × 90 cm³ coaxial or 3 × 20 cm³ planar HP Ge	50 keV–1 MeV or 20–120 keV	2.7% 1.5%	Various [9]

1979	Balloon	Univ. Durham	86 cm^3 HP Ge in LN$_2$ dewar	—	—	—	[10,11]
1979–80	HEAO-C Satellite	JPL	4 × 100 cm^3 HP Ge in CH$_4$/NH$_3$ cryostat	50 keV–10 MeV	~5%	Sun, SS433, gal. centre, Crab Nebula, interstellar medium [12–16]	
1980	Balloon	MPI/AIT	6 × 13 cm^3 HP Ge in LN$_2$ dewar	20–150 keV	~5%	Her X-1, Sco X-1 [17,18]	

[a] First balloon flight of Ge(Li) detector.
[b] First satellite-borne Ge(Li) detector.
[c] First satellite HP Ge detector.

Abbreviations: AIT, Astronomisches Institut, Tübingen; GSFC, Goddard Space Flight Center; JPL, Jet Propulsion Laboratory; LN$_2$, liquid nitrogen; MPI, Max Planck Institut, Garching; USAF, United States Air Force.

References: [1] Jacobsen (1968); [2] Imhof *et al.* (1974); [3] Nakono *et al.* (1974); [4] Jacobsen *et al.* (1975); [5] Ling *et al.* (1979); [6] Leventhal *et al.* (1977); [7] Leventhal *et al.* (1978); [8] Nakano *et al.* (1980); [9] Paciesas *et al.* (1983); [10] Ayre *et al.* (1981); [11] Owens (1985); [12] Mahoney *et al.* (1980); [13] Mahoney *et al.* (1981); [14] Riegler *et al.* (1985); [15] Mahoney *et al.* (1984); [16] Ling *et al.* (1983); [17] Proctor *et al.* (1982); [18] Jain *et al.* (1984).

4.5   HgI$_2$ detectors

The one serious disadvantage of both silicon and germanium X-ray spectrometers is that they must be cooled. Cryostats, and the other coolers now under investigation (section 4.3), tend to be large or heavy or power-hungry or short-lived; volume, mass, power and detector lifetime are always critically important in the design of X-ray astronomy experiments.

Since the early 1970s, interest has grown in the use of high-Z, large band gap, low leakage current semiconductors as room temperature radiation detectors. The first papers describing the γ-ray response of HgI$_2$, for example, appeared in 1972 (Malm, 1972). As one of the steps towards combining the operational simplicity of a proportional or scintillation counter with the energy resolution of silicon or germanium, a number of theoretical studies (Miller, 1972; Yee *et al.*, 1976; Armantrout *et al.*, 1977), working from known material properties, identified likely materials for further investigation. Of these, GaAs (Eberhardt *et al.*, 1971), CdS (Burger *et al.*, 1983), CdTe and HgI$_2$ have received the most subsequent attention, with regular international workshops devoted to the last two materials (Scheiber, 1978; Schieber and van den Berg, 1983). Cuzin (1987) has recently reviewed the status and ground-based applications of room temperature detectors, of which mercuric iodide is the only one so far used in X-ray astronomy.

The fabrication of large-area (~ 1 cm^2) mercuric iodide radiation detectors, described in detail by Schieber (1977) and Lamonds (1983), is rather more difficult than for silicon or germanium. As production techniques have improved over the past decade, the best experimental value for the Fano factor of HgI$_2$ has decreased steadily from 0.51 in the mid-1970s, to 0.19 in 1981 (Ricker *et al.*, 1982), to 0.1 at the present time (Iwanczyk *et al.*, 1985). The last value, obtained from X-ray measurements on a very small (1 mm^2) detector, is close to the theoretical value of 0.08 (Alig *et al.*, 1980).

The still imperfect state of development of HgI$_2$ in part reflects the relatively low economic importance of the compound in non-detector applications and in part arises from adverse properties of the material itself. Mercuric iodide undergoes a reversible phase transition (from tetragonal to orthorhombic crystal structure) at a temperature of 127°C which prevents crystals being grown from the melt. It is a soft material with a low melting point (Table 4.1) and high vapour pressure in the solid phase (0.03 Torr at 100°C, extrapolating to 2×10^{-5} Torr at 20°C; Weast, 1985). This last property can complicate the evaporation of metal contacts during detector manufacture (Scheiber, 1977) and would require the use of encapsulated crystals in any satellite experiment. The very binary nature of the compound is a problem, in that any departure from exact stoichiometry (an excess of mercury or iodine) seriously affects the electron transport properties of the material (Tadjine *et al.*, 1983). Trapping lengths (eq. 4.2, Table 4.1) in even the best mercuric iodide crystals remain short by comparison with silicon or germanium. Polarisation, the build-up within the crystal of a space charge field due to trapped carriers, can act to diminish the bias field and so degrade spectroscopic performance with time. Because of charge trapping, mercuric iodide is truly a room temperature detector. Cooling below $-20°C$ serves only to promote charge trapping and thus generates low-charge tails on any soft X-ray pulse height distribution (Ricker *et al.*, 1982).

Despite these problems mercuric iodide detectors, unencumbered by cryostats, remain attractive for X-ray astronomy. HgI₂ has already supplanted HP Ge in one 20–100 keV balloon payload and has been proposed as a 'replacement' for Si(Li) in the 1–20 keV band.

Fig. 4.11a represents the first use of mercuric iodide in X-ray astronomy. The hard X-ray spectrum of Cyg X-1 was measured by researchers at the Center for Space Research, MIT, during a balloon flight in June, 1980 (Ogawara *et al.*, 1982). The detector consisted of 11 HgI₂ crystals mounted in a cylindrical charged particle shield fabricated from NaI (see section 5.2.2.1). The energy resolution of the individual detector elements varied from 3.3–10% at 60 keV.

As pointed out by Vallerga *et al.* (1982, 1983), the high effective atomic number of mercuric iodide automatically renders it a 'low-background' hard X-ray detection medium. Compared to HP germanium, a thinner HgI₂ crystal is required (eq. 4.5, fig. 4.2) to achieve a given X-ray quantum efficiency at (say) 100 keV. This results in a reduction of background components (such as those due to neutrons and high-energy γ-rays) which scale with detector volume. The 'low-background' advantage of using a high-Z detector is maximised if the detector is shielded by a high-Z scintillator. Fig. 4.11b shows a prototype detector developed at MIT in which two 0.8 cm² HgI₂ crystals, fabricated by EG&G Inc., are enclosed by a 7.5 cm diameter bismuth germanate (BGO: $Bi_4Ge_3O_{12}$; Nestor and Huang, 1975) shield. The linear absorption coefficient of BGO is even greater than that of CsI (section 5.2) over the entire 10 keV–1 MeV energy range (Vallerga *et al.*, 1982). The top detector in the back-to-back pair had a depletion depth $d = 0.39$ mm which, together with the aluminium window thickness, defined an X-ray bandpass 20–100 keV. The second, anti-coincidence, crystal was 1 mm thick. Balloon flights of this instrument

Fig. 4.11. (a) Spectrum of black hole candidate Cygnus X-1. The high state upper limits labelled 'MIT 6/80' were obtained using a balloon-borne array of 11 HgI$_2$ detectors (Ogawara *et al.*, 1982). (*Reprinted by permission from* Nature, *vol. 295, p. 675.* © *1982 Macmillan Magazines Ltd.*) (b) Bismuth germanate shielded HgI$_2$ detector (Vallerga *et al.*, 1982). X-rays enter through the 0.5 mm aluminium window at the left hand side. (*Courtesy G. Ricker.* © *1982 IEEE.*)

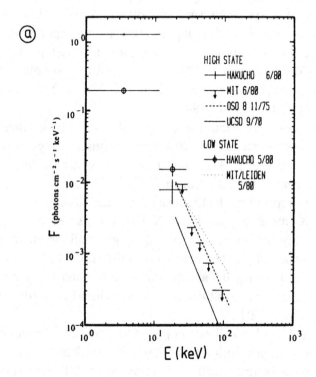

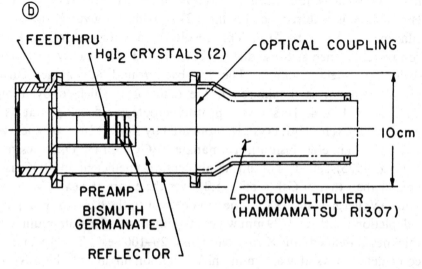

(Vallerga *et al.*, 1983) produced background count rates in the 40–80 keV band

$$B_i = 4.2 \times 10^{-5} \text{ counts/cm}^2 \text{s keV}$$

in good agreement with pre-flight Monte Carlo simulations and about a factor two lower than the 'ultimate' background level of an optimised balloon-borne scintillation counter (section 5.2.2.1).

In order to turn this background advantage into a gain in point source sensitivity (eq. 1.1), Ricker *et al.* (1983) have proposed using a hard X-ray flux concentrator (section 1.4) – consisting of nested LiF – and graphite-lined paraboloids to 'focus' 40–80 keV X-rays on to a small HgI₂ detector of the kind illustrated in fig. 4.11a. The alternative approach – producing much larger mercuric iodide crystals without loss of energy resolution – is not practical at the present time.

Arrays of mercuric iodide crystals have recently been considered as focal plane detectors for the XMM mission (Di Cocco *et al.*, 1985; Briel *et al.*, 1988) as an alternative to the Si(Li) mosaics described in section 4.3. The 60 keV response of an 8 × 8 element HgI₂ array has been described by Ortale *et al.* (1983) at EG&G. In this prototype, orthogonal sets of linear contacts were produced on the front and back surfaces of a crystal slice by photolithography. 'Row-and-column' logic at the 16 amplifier outputs was used to determine the position of the active 1.8 × 1.8 mm² element. Individual elements were spaced 0.2 mm apart.

Groups at Bologna and Strasbourg (Di Cocco *et al.*, 1985) have begun development, for space applications, of soft X-ray detector arrays with much smaller pixel sizes (0.2 × 0.2 mm²). The 'window effect' (section 4.2.3) in mercuric iodide detectors (at least those with carbon contacts) is much less serious than in germanium, so that kilovolt X-rays can be detected with good energy resolution (Dabrowski *et al.*, 1981). Effective dead layer thicknesses of only 0.1–0.2 μm are observed in HgI₂, so that the bandpass of any such array would be determined by the nature of its UV and environmental shields.

The best energy resolutions reported so far for HgI₂ (fig. 4.3) have been obtained with room temperature detectors coupled to preamplifiers cooled by liquid nitrogen, a significant step away, in satellite terms, from the promise of a cryostat-free semiconductor detector. Modest thermoelectric or passive (radiative) cooling of both FET and detector crystal would lead to an energy resolution

$$\Delta E = 175 \text{ eV}$$

independent of X-ray energy in the 1–10 keV band (Iwanczyk *et al.*, 1985; Briel *et al.*, 1988).

Finally, the long-term stability of mercuric iodide under particle bombardment is unknown. Ground-based measurements (Becchetti *et al.*, 1983), however, indicate that HgI_2 has greater resistance to neutron damage than silicon.

4.6   CCDs

The first charge coupled devices were produced at Bell Telephone Laboratories in 1970 (Boyle and Smith, 1970). The potential of CCDs as high resolution imaging detectors was realised very quickly thereafter (Gordon, 1972) and the first demonstration of X-ray sensitivity (Koppel, 1977; Peckerar *et al.*, 1977) was followed almost immediately by the first detection of single 5.9 and 22.4 keV photons (Catura and Smithson, 1979). It became apparent to many X-ray astronomers that the cooled CCD, with certain easily identified developments (Schwartz *et al.*, 1979), could become the 'all-singing, all-dancing' detector (section 1.6), made flesh (or rather silicon) at last.

Whereas the segmentation of Si(Li) diodes (fig. 4.8) produces positional sensitivity on millimetre scales at best, and while signal location by charge division between two or four electrodes in a silicon crystal (Laegsgaard, 1979) is ruled out at soft X-ray energies on signal-to-noise grounds, the CCD formats described in the next section naturally offer the energy resolution of silicon on linear scales of $\sim 20\,\mu m$.

The past decade, in which CCDs have completely revolutionised optical astronomy (Mackay, 1986; Janesick and Blouke, 1987), has seen much of the early optimism fulfilled. Those areas of CCD X-ray detector research which were within reach of the average X-ray astronomy group – device modelling, optimisation of the signal processing chain – are now highly

Table 4.5. *Major manufacturers of scientific CCDs*

English Electric Valve Company (EEV)
Waterhouse Lane, Chelmsford CM1 2QU, Essex, England

Texas Instruments Inc. (TI)
13500 North Central Expressway, Dallas, Texas 75265, USA

Tektronix Inc.
PO Box 500, Beaverton, Oregon 97077, USA

Thomson CSF
Division Tubes Electroniques, 38 Rue Vauthier,
92102 Boulogne-Billancourt CEDEX, France

perfected. Those developments, identified right at the start of the CCD story – deep depletion, back illumination, mosaicing (section 4.6.2) – which concerned device technology rather than device operation are, however, in a much less complete state. Scientific charge coupled devices, like micro-channel plates (Chapter 3), are based on a mass-market commodity – TV sensors in the CCD case – whose production it is costly to perturb for the sake of a few specialised units. Front line research into CCDs for X-ray astronomy thus normally requires the close cooperation of a device manufacturer (Table 4.5) and relatively high levels of funding. Conversely, if a manufacturer withdraws from the marketplace (as RCA did in late 1985), dependent CCD research groups (Lockheed Palo Alto and MSSL in the case of RCA; Walton *et al.*, 1985) are severely disadvantaged.

Active CCD consortia in X-ray and XUV astronomy now include:

(i) Pennsylvania State University, JPL and MIT, who are working, in collaboration with Texas Instruments, on the definition phase of the AXAF CCD imaging spectrometer (ACIS; section 4.6.3).

(ii) The Leicester X-ray Astronomy Group and ESA's Space Science Department, who are developing, in collaboration with EEV, CCD arrays for the Western European/Soviet JET–X experiment (section 1.5) and for XMM.

(iii) Lockheed Palo Alto research laboratories, Johns Hopkins and MSSL, who, with Tektronix, have embarked on a programme to enhance the UV and XUV response of CCDs (Stern *et al.*, 1987).

One of the few research groups with the ability to fabricate its own CCDs is the US Naval Research Laboratory, which is investigating the use of back illuminated devices in the UV and XUV (D. Michels, unpublished paper, 1986).

4.6.1 *CCD format and operation*

A charge coupled device is essentially an array of MOS (metal-oxide-silicon) capacitors formed by the deposition of parallel linear electrodes on to the oxidised surface of a silicon wafer (fig. 4.1e). Electrons are the majority carriers in most CCDs. Typical 525/625 line compatible TV format CCDs have active areas of about 1 cm^2: the largest specialised CCD fabricated to date is the Tektronix 2048 × 2048 pixel array (Janesick and Blouke, 1987). This device – the world's largest integrated circuit – is both expensive ($\sim$ US\$80 000 each) and so far unproven in single photon counting (section 4.6.2). Although linear CCDs for use in ultra-fast signal processing have been constructed from GaAs (Cresswell *et al.*, 1986) the X-ray and optical response of such devices is at present unknown. The prospect of high speed (electron mobilities are a factor six higher in GaAs than in Si; Cuzin, 1987), room temperature (section 4.5) CCD operation may, however, attract researchers in the near future.

Most CCDs currently under investigation for X-ray astronomy are three-phase, frame transfer devices, whose operation is illustrated in figs. 4.12 and 4.13. Application of a positive voltage V_G to any one of the electrodes running horizontally in fig. 4.13 establishes a depletion region beneath it, into which particle, light or X-ray induced signal charges will accumulate as long as the voltage is maintained. If, after an appropriate integration time ('frame time'), the voltage on that electrode is reduced and the voltage on its neighbour raised, the accumulated charge will be transferred ('coupled') to the new depletion region. The frame time sets the detector time resolution in normal operation.

In a three-phase CCD the electrodes, consisting of overlapping layers of polycrystalline silicon ('polysilicon'), are connected cyclically in triplets. Three electrodes thus constitute the width of a resolution element for imaging, centred on the electrode biased high during integration. The other pixel dimension is determined by the spacing of heavily doped p-type implants ('channel stops') which prevent the diffusion of charge along an electrode's length. TV format devices usually have square pixels 15–30 μm on a side, and are divided into two sections: a photosensitive image section and a store section, the latter covered by an opaque shield (fig. 4.13). After integration, the contents of the image section are rapidly transferred, row by row, into the store, from which they can be read out without undergoing further illumination. In all astronomical applications, CCDs without light shields are preferred and images are accumulated over the entire array ('full frame' operation). The problem of image smear due to the accumulation of photons during readout is dealt with in optical astronomy by use of a mechanical shutter and in X-ray astronomy by arranging for the readout time to be a small fraction (~ 0.1) of the integration time.

In any case, the last row of store pixels always transfers to a serial line readout section, which is essentially a one-dimensional CCD shift register terminated at one end by an output amplifier (fig. 4.13). If the frame time is chosen such that the probability of collecting more than one photon per pixel per frame is small, the amplifier output pulse train contains both position and energy information. Each voltage pulse is uniquely associated with a particular CCD pixel simply by virtue of its place in the train, while the amplitude of any pulse is proportional to X-ray energy. For random

photon arrival at a mean rate F photons per pixel per frame time the
probability of two or more events within one resolution element is

$$P(2, 3, ...) = 1 - P(0) - P(1) = 1 - (F + 1)\exp(-F) \qquad (4.8a)$$

Fig. 4.12. Parallel charge transfer clock pulses in a three-phase CCD (EEV
CCD Imaging, 1987). The top three diagrams show the location of a signal
charge packet at times t_1, t_2, t_3 in the clock sequences φ_1, φ_2, φ_3 applied to
electrode groups 1, 2, 3. Typically, $V_G = 10\text{--}12$ V. (*Courtesy EEV Ltd.*)

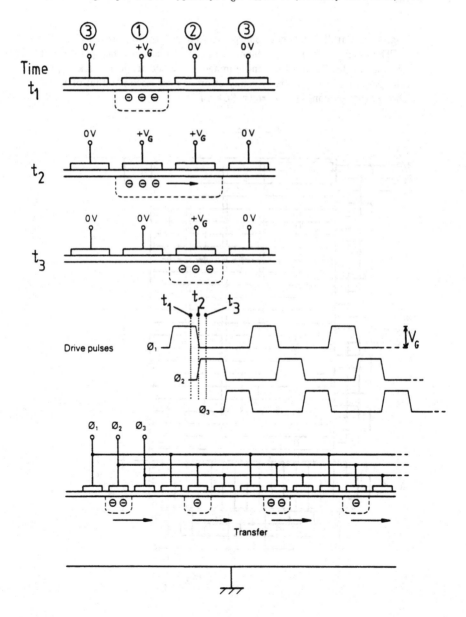

To restrict event confusion to the 1% level ($P(2, 3, ...) < 0.01$) then implies $F < 0.15$ photons per pixel per frame, or rather less if on-chip binning is employed (section 4.6.2). If such count rate constraints are not met, pulse amplitude becomes a measure of the source intensity only.

Pixel readout rates in X-ray evaluation are around 50 kHz, giving readout times of 2–5 s in standard devices. Pixel readout rate is limited by the need to minimise amplifer noise (below), and hence the bandwidth of the signal processing chain. Image smear, event confusion and noise

Fig. 4.13. Parallel and serial charge transfer in a frame transfer CCD (EEV CCD Imaging, 1987). This schematic diagram illustrates the operation of a 5×4 pixel array. Iφ, Sφ and Rφ denote image, store and readout clock sequences. The periods of I and S are typically of order microseconds; that of R, much shorter (nanoseconds). (*Courtesy EEV Ltd.*)

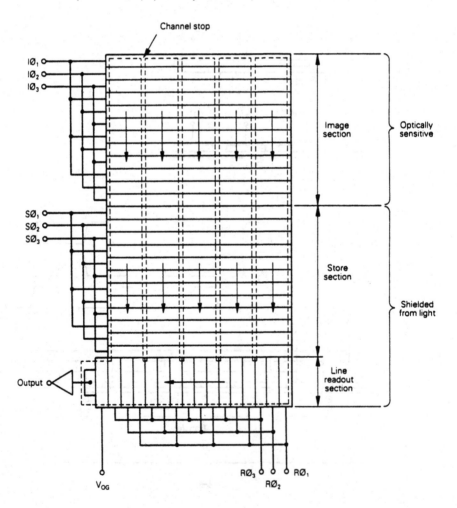

Fig. 4.14. Timing tradeoffs in a full-frame CCD camera counting single X-ray photons. The optimum combination of readout time and frame (integration) time (~ 2 s, 20 s (circle) for a ~ 1 cm² device operating at a 50 kHz pixel readout rate) lies at the bottom vertex of the triangle in parameter space enclosed by the lines *a*, *b*, *c*. Smaller values of readout time (to the left of line *a*) entail a higher noise bandwidth and hence degraded energy resolution. Readout time/frame time ratios less than about 0.1 (below line *b*) give rise to unacceptable image smear, due to the accumulation of photons during readout. Values of frame time above a limit (line *c*) determined by source intensity yield an unacceptably high probability of collecting two or more photons in the same pixel (eq. 4.8). For bright sources, the triangle may disappear completely, and alternative methods of CCD readout may have to be employed.

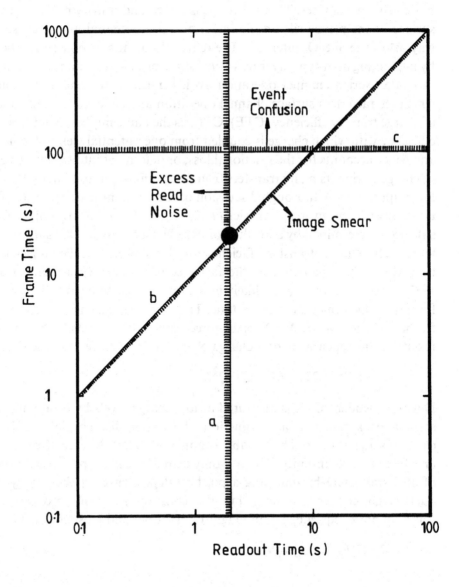

bandwidth problems together constrain the choice of integration and readout times for a CCD X-ray detector in the manner shown in fig. 4.14.

The CCD format illustrated in figs. 4.12 and 4.13 is by no means unique. Virtual phase CCDs, produced by Texas Instruments (Stern *et al.*, 1983) are essentailly two-phase devices in which one set of polysilicon electrodes (gates') is replaced by a series of ion implants. The single gate design of virtual phase CCDs leads, in principle, to superior low energy X-ray and optical quantum efficiencies by removing some of the dead layers (section 4.2.2) inherent to the three-phase 'front illuminated' design.

All CCDs now used in astronomy, irrespective of electrode structure, are buried channel devices. That is, charge is stored and transferred, not at the surface of the p-type silicon (fig. 4.1e) but in a potential minimum some way below the Si-SiO$_2$ interface. The required potential profile is produced by implanting an n-type layer between the silicon substrate and its overcoat of oxide. Buried channel operation avoids much of the charge trapping which characterises surface channel operation and so leads to high values of charge transfer efficiency (CTE). CTE, as the name implies, describes the efficiency with which charge is coupled from one potential well to the next, and hence accounts for the fractional loss, or deferment into trailing pixels, of charge during its many transfers from the point of photon interaction to the output node. A thorough discussion of the mechanisms responsible for non-unity CTE is given by Janesick *et al.*, 1985b, 1987). The overall CTE of a device is determined by a product of parallel (suffix p) and serial (suffix s) terms. The charge transfer efficiency of the slower, parallel register is generally higher than that of the faster serial register (Epperson *et al.*, 1987), since electrons are less likely in the former case to be locally trapped for times longer than the transfer time. Transfer of signal charge S from the furthest corner of an $N \times N$ pixel three-phase array to its output node results in the appearance of a charge S' in that pixel at readout, where:

$$S' = CTE_s^{3N}CTE_p^{3N}S \sim CTE_s^{3N}S$$

Device-dependent CTE is measured at low charge levels by observing the shift in peak position, as a function of transfer distance, of the 1620 electron signal produced by the absorption of a 5.9 keV Mn Kα X-ray from an ^{55}Fe source. Non-unity CTE not only complicates the spectroscopic use of an X-ray CCD by imposing a position dependence on the energy-to-pulse height conversion factor, but also constitutes a position dependent source of noise (eqs. 4.6, 4.7) of magnitude (Carnes and Kosonocky, 1972):

$$R = (2[1 - CTE_s]N_T S)^{\frac{1}{2}}$$

This equation simply quantifies the fluctuation in the number of electrons lost from a signal batch of S electrons, after N_T serial transfers. Values of serial CTE for the best available CCDs now exceed 0.999995, so that this noise source can be made negligible compared to others. It should be noted, however, that such high values can generally only be achieved at temperatures above 170 K, due to the so-called 'fat zero' effect of thermally generated electrons in filling trapping sites (Janesick *et al.*, 1987). As the thermal electron population decreases with temperature, these trapping sites act to reduce the signal electron population during transfers.

In CCD work, the terms A and R of eq. (4.7) are lumped together into a single detector parameter – the read noise – which in modern devices has values of order ten electrons rms at 50 kHz readout rates. Such low values are, of course, essential if CCDs are to be used in high-resolution spectroscopy (section 4.2.3). It is possible that, in the future, optimisation of the on-chip FET could reduce this level to one electron or less (Janesick *et al.*, 1987). One should of course note that these impressive equivalent noise charges appear on node capacitances of order 0.1 pF, so that the noise voltage implied by one electron noise is a respectable 1.6 μV. Detailed accounts of cooled amplifier design and noise components are given by Bailey *et al.* (1983), Mackay (1986) and Janesick *et al.* (1987). Hopkinson and Lumb (1982) describe the variants of the correlated double sampling technique used to attenuate both low frequency ($1/f$) noise and the 'reset' noise which arises from the restoration of the node potential to a reference level after every pixel has been clocked out. Samples of the voltage level are taken (by integration for a fixed time or using peak sample-and-hold circuitry) both before and after the signal charge is transferred to the output node, so eliminating the reference level and its variation from pixel to pixel. As noted above, read noise increases with pixel readout rate reaching about 50 electrons rms at 1 MHz (Damerell, 1984). Read noise generally has a minimum value in the temperature range 150–180 K (Chowanietz, 1986; Lumb and Holland, 1988a).

Read noise, together with system gain in final ADC channels per photoelectron, may be measured by the optical mean-and-variance or photon transfer (Mortara and Fowler, 1981; Janesick *et al.*, 1987) technique (fig. 4.15). Alternatively, the system gain may be measured by recording the pulse heights of a series of X-ray emission lines and using the value of w for silicon appropriate to the device temperature (section 4.2.1) to convert X-ray energies to mean signal levels (Lumb and Holland, 1988a).

4.6.2 *Developments for X-ray astronomy*

4.6.2.1 *Deep depletion*

TV format CCDs are usually fabricated from low-resistivity epitaxial silicon ($\rho_s \sim 10\text{--}25\ \Omega\,\text{cm}$; $N_{AD} \sim 10^{15}\,\text{cm}^{-3}$ acceptors) grown on a more highly doped silicon substrate ($N_{AD} \sim 10^{18}\,\text{cm}^{-3}$). The 1 μm thick electrode layer (fig. 4.1e) overlays a $\sim 20\,\mu\text{m}$ epitaxial region, of which the first 3–10 μm are depleted and the remainder is 'field-free', and a substrate several hundred microns thick. These layered structures possess rather limited X-ray sensitivity. The electrodes efficiently absorb sub-kilovolt

Fig. 4.15. Measurements of EEV P8603 CCD read noise and electronic system gain g channels per electron by the mean-and-variance technique (Lumb and Holland, 1988a). If a small area of a CCD is illuminated by an optical intensity corresponding to mean ADC output channel S, the channel variance σ_s^2 will be the sum of contributions from the signal variation itself ($= S$) and from the read noise. Expressed in channels (and with an astronomer's disregard for dimensional correctness), $\sigma_s^2 = gS + g^2\,(A^2 + R^2)$. Straight line fitting gives $g = 0.522$ channels/electron, in good agreement with the value obtained from independent X-ray measurements, and a read noise of 9.6 electrons rms. (*Courtesy D. H. Lumb. © 1988 IEEE.*)

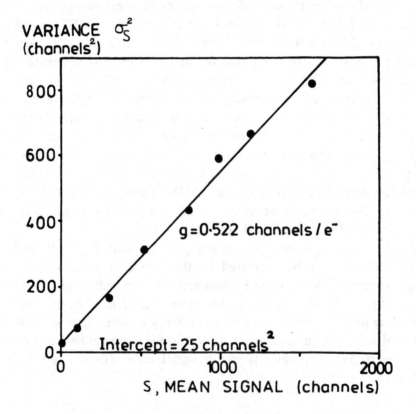

VARIANCE σ_S^2
(channels²)

800 —

600 —

400 — $g = 0.522$ channels / e⁻

200 —

0 —   Intercept = 25 channels²

0   1000   2000

S , MEAN SIGNAL (channels)

photons, while the depletion region is transparent to energies above about 6 keV (fig. 4.2).

Improved high-energy response – highly desirable in astrophysics for the observation of iron line complexes around 6.7 keV (Wells *et al.*, 1985) – requires the fabrication of 'deeply depleted' CCDs on silicon substrates of higher resistivity ρ_s (i.e. with lower impurity levels). Depletion depth d varies with the square root of resistivity according to eq. (4.3).

The first attempt to fabricate such devices was, within the limits of its application, highly successful. Peckerar *et al.* (1981) reported 98% quantum efficiency and good image response for 8 keV Cu K X-rays incident on a 200 element linear array manufactured from 10 kΩ cm silicon. It has,

Fig. 4.16. Progress in CCD energy resolution ΔE and quantum efficiency Q at an X-ray energy of 5.9 keV. Device type and references: [1] EEV MA328; Lumb and Hopkinson (1983). [2] Fairchild 221; Griffiths (1985). [3] RCA 501 EX002; Walton *et al.* (1984). [4] TI 800 × 800; Janesick *et al.* (1985a). [5] EEV P8603; Lumb *et al.* (1985). [6] Bell Northern Research W62A; Griffiths (1985). [7] RCA 53834X0; Walton *et al.* (1984). [8] EEV P8600HR (Batch 1); Schwartz *et al.* (1985). [9] EEV P8600HR (Batch 2); Chowanietz *et al.* (1985). [10] TI 4849; Garmire *et al.* (1985). Areas in the parameter plane above the broken horizontal lines are forbidden to devices with resistivities less than the stated values. The area to the left of the vertical line is excluded by Fano-factor statistics (eq. 4.7). Whereas the point I represents the ideal CCD detector, the point G represents the current goal of device development programmes directed towards both AXAF and XMM missions.

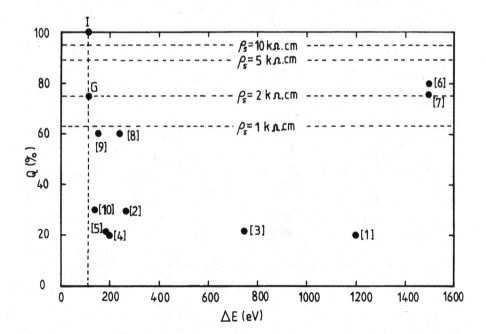

unfortunately, proved much more difficult to adapt the CCD manufacturing technology to high-resistivity silicon without sacrificing spectroscopic performance, through degraded read noise and/or CTE. Fig. 4.16 records recent progress towards the goal of simultaneously high efficiency and good energy resolution at an X-ray energy of 5.9 keV. Griffiths (1985), Lumb *et al.* (1985) and Walton *et al.* (1985) describe in detail the shortcomings of early experimental devices.

The best overall performance reported to date is that achieved at Leicester with 30 μm depleted, 578 × 385 pixel EEV P8600HR (high resistivity) CCDs manufactured on 1 kΩ cm silicon (Chowanietz *et al.*, 1985). Fig. 4.17 compares the 1–10 keV efficiencies of standard and high-resisti-

Fig. 4.17. Quantum efficiency as a function of X-ray energy for EEV P8607 ($d = 3$ μm) and P8600HR ($d = 30$ μm) CCDs (Chowanietz *et al.*, 1985). Individual symbols: measured values; curves: theoretical estimates. (*Courtesy D. H. Lumb.*)

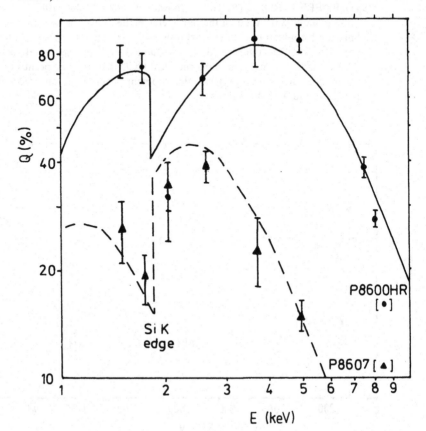

Fig. 4.18. CCD pulse height distributions (single pixel X-ray events) (Chowa-nietz *et al.*, 1985). (a) EEV P8607. (b) EEV P8600HR. (*Courtesy D. H. Lumb.*)

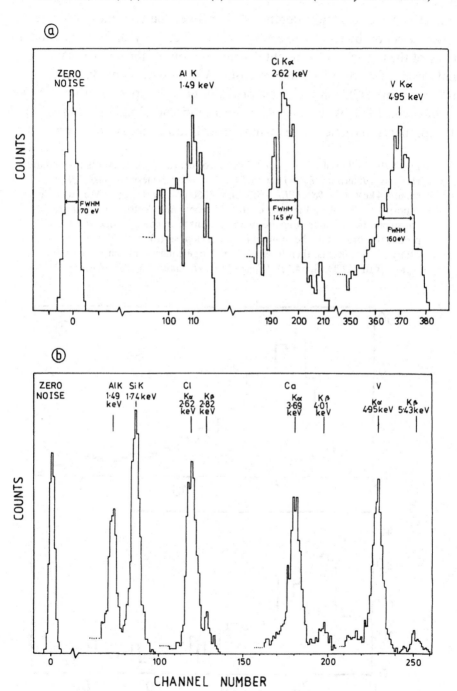

vity CCDs while fig. 4.18 compares pulse height distributions from the two devices.

Increasing the depletion depth also improves the efficiency of charged particle rejection by energy selection (section 4.2.4), by shifting the modal energy of the particle pulse height distribution out of the 0.5–10 keV X-ray band desired for most satellite missions. Measurements made with standard ($d \sim 5\,\mu$m) CCDs show a most probable energy deposit of about 5 keV (Bailey et al., 1983; Walton et al., 1985; Lumb and Holland, 1988a). Fig. 4.19 compares particle pulse height distributions measured with both

Fig. 4.19. CCD pulse height distributions of charge particle events (Lumb and Holland, 1988a). (a) EEV P8603 ($d = 5\,\mu$m). Modal energy deposit 5.0 keV, mean 6.4 keV. (© *1988 IEEE*.) (b) EEV P8600HR ($d = 30\,\mu$m). Modal energy deposit 17 keV. Using the 'rule of thumb' secondary electron creation rate 80/ μm silicon track length, together with an ionisation energy value $w = 3.65$ eV, produces estimates of the minimum energy deposit (a) 1.5 keV, (b) 8.8 keV. Many particle events, therefore, feature charge collection from the field-free region of the CCD (fig. 4.1e). (*Courtesy D. H. Lumb. © 1988 IEEE*.)

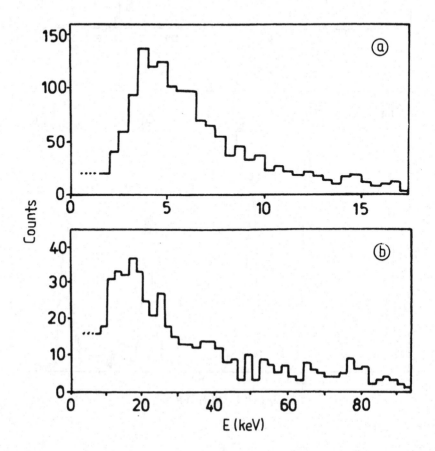

standard and deeply depleted CCDs. In the latter case, the modal energy deposit is 17 keV (Lumb and Holland, 1988a).

Background events in deeply depleted CCDs may also be recognised on the basis of morphology. In deeply depleted devices, most X-ray absorption occurs within the depletion region and so gives rise to 'single pixel' events. Particles, on the other hand, deposit energy in both field-free and depletion regions. Because charge liberated in the field-free region is liable to spread into more than one pixel before collection, particle events may occupy many (up to 10×10; Wells *et al.*, 1985) adjacent pixels. Essentially complete rejection of laboratory particle background has been claimed for pulse height and spread event rejection techniques used in conjunction (Lumb and Holland, 1988a).

A detailed analysis of charge diffusion in the field-free region of a CCD is given by Hopkinson (1983, 1987). The latter paper also considers the implications of 'on-chip binning' (next section) on CCD quantum efficiency and energy resolution. Janesick *et al.* (1985b) address the same electron transport problems by means of a Monte Carlo model.

4.6.2.2 *On-chip binning*

The spatial resolution offered by small pixel CCDs is by no means well-matched to telescope resolution in many of the satellite missions now being planned. In the case of ESA's XMM (section 1.5) the half energy width of the telescope blur circle has a target value of 30″, equal to about 1 mm in the focal plane. A CCD operated in standard full frame mode would therefore linearly oversample the telescope resolution by about a factor of 50. A technique used in optical astronomy (Kramm and Keller, 1985) – on-chip binning: the addition, by suitable manipulation of the CCD clocks, of adjacent CCD elements to create larger 'superpixels' – has been investigated as a means of eliminating gross oversampling and drastically reducing the amount of CCD data bits produced per frame.

Pixel manipulation in X-ray astronomy implies a tradeoff between quantum efficiency and energy resolution (Hopkinson, 1987) with complications of background noise and count rate limitations thrown in for good measure. On the plus side:

(i) The detector time resolution (equal to the integration time and necessarily proportional to the readout time) is improved by a factor M^2 if $M \times M$ pixels are binned together.

(ii) The detector quantum efficiency increases with M (rapidly in standard devices, more slowly in deeply depleted CCDs; Lumb *et al.*, 1987) as signal charge which may have diffused into several pixels is re-integrated into one packet.

On the debit side, however:

(i) Acceptable count rates for single photon counting decrease with M. If $M \times M$ pixels are binned together, the probability that more than one photon is counted per superpixel per frame becomes

$$P(2, 3, ...) = M^2 F \exp(-F)[1 - \exp(-(M^2 - 1)F)] \qquad (4.8b)$$

Here, as in eq. (4.8a), F is the source count rate per CCD pixel per frame. Table 4.6 shows how the value of F consistent with 1% event confusion decreases with increasing superpixel size. For a putative CCD camera in the focus of an XMM mirror module, the brightest source which can be examined in full frame mode decreases from about 0.033 to 0.023 times the intensity of the Crab Nebula as M increases from one to four. These calculations bring out the important count rate limitations of CCD X-ray spectrometers operated in the full frame mode described in figs. 4.12 and 4.13. The implication of Table 4.6 is that a CCD camera on XMM would not be able to point at the 40 brightest celestial X-ray sources, unless non-standard X-ray operating modes, such as using continuous readout to translate CCD rows into time bins (Briel *et al.*, 1988) or reverting to the use of a shielded store section (fig. 4.13), were employed. These limitations would become even more serious if real time pattern recognition algorithms were used to reject background events on the basis of charge spreading (previous section).

(ii) Energy resolution is, in general, degraded by pixel binning. The

Table 4.6. *Variation of acceptable count rate* F *for single photon counting with superpixel size* M

Event confusion level 1% ($P(2,3,..) = 0.01$; eq. 4.8). Assumed CCD pixel size 20×20 microns. Frame time $20/M^2$ s. Count rate from Crab Nebula in focal plane of XMM mirror module $= 1000$ counts/s in 1.5 mm diameter spot (Briel *et al.*, 1988).

M	F (single photon limit) (counts per CCD pixel per frame)	F (Crab Nebula) (counts per CCD pixel per frame)
1	0.15	4.5
2	0.03	1.13
3	0.012	0.50
4	0.0065	0.283

summation of adjacent charge packets is an inherently noiseless process which nevertheless broadens pulse height distributions to low signal levels because of incomplete charge collection effects, especially in low-resistivity CCDs (Lumb and Holland, 1988b). In devices with relatively low CTE (~ 0.99992; Chowanietz, 1986) 2×2 binning may lead to a 'double peaked' response to a mono-energetic stimulus, depending on whether or not the bulk of the signal is in the leading or trailing column of the superpixel. In the former case, on-chip binning reunites the signal with its deferred charge; in the latter case it does not. It should be noted that the optimum CCD energy resolution plotted in fig. 4.3 corresponds not to summed pixel events, nor even to single pixel events, but to isolated pixel events where no significant charge (relative to the dark noise baseline) is found in any adjacent pixel. At 4.5 keV, accepting all single pixel events increases ΔE from 140 to 170 eV. Introducing 3×3 binning increases the fwhm energy resolution still further, to 250 eV (Lumb and Holland, 1988b).

(iii) Background rejection algorithms, operating on the basis of charge cloud size, become less efficient as M is increased. Lumb and Holland (1988a) have reported a fall in particle rejection efficiency from $\sim 99.9\%$ to 50% as the effective pixel size increased from 17 to 88 µm (an implied increase in particle count rate of 50 times).

4.6.2.3 *Back illumination*

Whereas the depletion depth d controls the high-energy X-ray response of a CCD, its low-energy response in normal, front illuminated operation (fig. 4.1e) is determined mainly by the electrode structure. Standard, three-phase devices are 'blind' in the wavelength range 4000 Å to 50 Å, while the Texas Instruments virtual phase devices discussed in section 4.6.1 have an improved ultraviolet threshold of ~ 1800 Å (Janesick *et al.*, 1987). Some work has also been done using phosphor converters (section 5.3.2) to improve the XUV response of standard CCDs.

The prospect of producing a single 'four decade' detector sensitive across the complete 1–10 000 Å wavelength range has motivated recent studies of back illuminated CCDs, in which the substrate and epitaxial layers of fig. 4.1e are thinned away to expose the depletion region for direct illumination. Thinning is achieved by a combination of mechanical abrasion and chemical etching (Lesser *et al.*, 1986). Since XUV light is absorbed within a few hundred angstroms of silicon, the treatment of the exposed surface is critical if high quantum efficiency is to be achieved (Janesick *et al.*, 1985a, 1986, 1987; Stern *et al.*, 1986, 1987). Bare silicon develops a $\sim 30\,\text{Å}$ native oxide layer in air. This layer may contain sufficient trapped positive charge to create a backside depletion region which captures signal electrons in preference to the buried channel at the front surface. Various techniques including flooding the device with UV light, deposition of a metallic monolayer – the so-called flash gate, developed at JPL – and ion implantation have been used to modify the charge collecting field near the exposed silicon surface. All these techniques have suffered from problems of reproducibility and stability. For example, Stern *et al.* (1987) report how the $584\,\text{Å}$ quantum efficiency of an experimental ion-implanted CCD decayed from an initially high level of 50% to 6% in only three-and-a-half hours. This (exponential) decay was attributed to the build up of a volatile contaminant (possibly water ice) on the cooled CCD.

For missions such as AXAF and XMM, the low-energy response of a CCD camera (section 4.6.3) is always likely to be limited by the thickness of its UV/optical shield. The goal of back illumination for X-ray CCDs, therefore, is the relatively modest one of shifting the detector cutoff down from about $0.7\,\text{keV}$ to the carbon K edge. This requires that any backside dead layers be reduced in thickness to about $500\,\text{Å}$. Rather more strict constraints apply if CCDs are to supplant microchannel plates in XUV and UV space astronomy (section 3.4). Comparisons between CCDs and MCPs for the Lyman mission (Stern *et al.*, 1986; Vallerga and Lampton, 1988) have tended to yield somewhat partisan results, although folding in practical considerations such as cooling, optical sensitivity and detector size does seem to shift the balance very definitely in the direction of channel plates. MCPs, moreover, retain the fundamental ability to count single photons throughout their bandpass, while CCDs must be operated, as in optical astronomy, as integrating detectors whenever the signal charge per photon and the detector read noise become comparable. This implies a low-energy, long wavelength limit for single photon counting with a CCD such that:

$$E/w \gtrsim 3A$$

Taking $A \sim 10$ electrons for current devices yields $E > 0.11\,\text{keV}$. The lowest energy X-ray line for which a pulse height distribution has been recorded experimentally is C K ($0.277\,\text{keV}$; Janesick *et al.*, 1985a).

4.6.3   *Flight instruments*

No CCD X-ray camera has yet flown on a satellite. Other than the sounding rocket flights proposed by the Penn State/JPL/MIT consortium

(Nousek *et al.*, 1987), the first flight opportunity may well be the JET-X experiment circa 1993, at least 16 years after the first measurements of X-ray sensitivity. Optical CCD cameras have, however, already been space qualified. The Halley Multicolour Camera on ESA's Giotto spacecraft (Keller *et al.*, 1987b) contained two Texas Instruments 390×292 CCD arrays, while the auroral imager on the Swedish Viking satellite (section 3.4.1) employed a CCD to read out a microchannel plate image intensifier.

In the longer term, it is likely that CCDs will perform a major role in both the AXAF and XMM observatory class missions (section 1.5). In neither case has a definitive instrument design yet emerged. The AXAF CCD imaging spectrometer is currently in a definition phase, having been descoped by NASA from its original size (Garmire *et al.*, 1985). Typical CCD pixel sizes are well-matched to the high-resolution AXAF optics (telescope parallax effects (eq. 2.11) have negligible impact on detector resolution in all solid state detectors). In the case of XMM, a CCD camera has been identified as the prime instrument in a so-called 'strawman' payload for further industrial study (Briel *et al.*, 1988). CCDs for both these and other missions face a number of common problems:

(i) Cooling. The optimum CCD operating temperature, ~ 180 K, is determined by CTE and read noise requirements (section 4.6.1). Here, the comments made in connection with Si(Li) detectors in section 4.3 apply. Passive cooling by radiators may be an option in certain deep space orbits being considered for XMM (Briel *et al.*, 1988).

(ii) Focal plane coverage. A single TV format CCD array would fill only a small fraction of either the AXAF (18 cm diameter) or XMM (7 cm diameter) fields-of-view. The use of a single large custom-built chip such as the 55×55 mm² Tektronix 2048×2048 array (Janesick and Blouke, 1987) is unlikely to provide a satisfactory alternative, since the 20-fold larger number of pixels would require a corresponding increase in readout time. Detector time resolution and single photon counting capability would then be severely degraded (see fig. 4.14 and eqs. 4.8a, b). Secondly, as CCDs become larger, their charge transfer efficiency must improve still further. Fig. 4.20 shows the results of a Monte Carlo simulation of the ^{55}Fe response of CCDs with different sizes and CTE values.

The more likely solution to the coverage problem is, as with Si(Li) and HgI₂, to form a close packed mosaic (2×2 or 3×3) of small detectors. Ellul *et al.* (1984) have described the design of 'buttable' CCD arrays which may be connected end-to-end with less than 100 µm dead spaces between them. More conservative approaches, mounting each CCD on its own individual carrier for testing before integration into the mosaic, might involve dead spaces several millimetres wide. Unlike large single CCDs, mosaics offer the safeguard of redundancy in the event of chip failure and can be configured 'vertically' to match the curvature of the telescope focal surface.

(iii) Radiation damage. Little data exists regarding the long-term

Fig. 4.20. Simulated ^{55}Fe X-ray pulse height distribution as a function of CCD array size and charge transfer efficiency (CTE) (Janesick *et al.*, 1987). (*Courtesy J. R. Janesick.*)

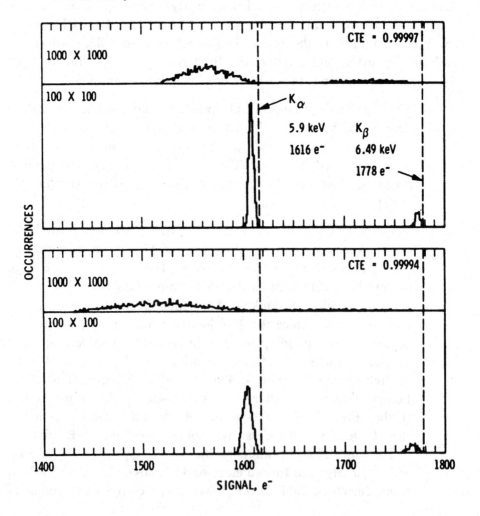

stability of cooled CCDs under particle bombardment. The accumulation of positive charge in CCD oxide layers (like the polarisation effect in HgI_2; section 4.5) may cause changes in the voltage across the depletion region, while Si lattice damage caused by neutron and proton interactions may result in increased trapping of charge carriers and hence may reduce CTE and increase dark current in the vicinity of the defects (Damerell, 1984).

A conceptual view of a CCD detector for XMM is shown in fig. 4.21.

Fig. 4.21. Schematic cross-section view of possible CCD camera for XMM. *Key*: 1, light shield; 2, inner thermal baffle; 3, outer thermal baffle; 4, ion pump port; 5, cold finger; 6, spacecraft thermal interface; 7, 3×3 CCD mosaic; 8, electrical feedthroughs; 9, spacecraft electrical interface. (*Courtesy D. H. Lumb.*)

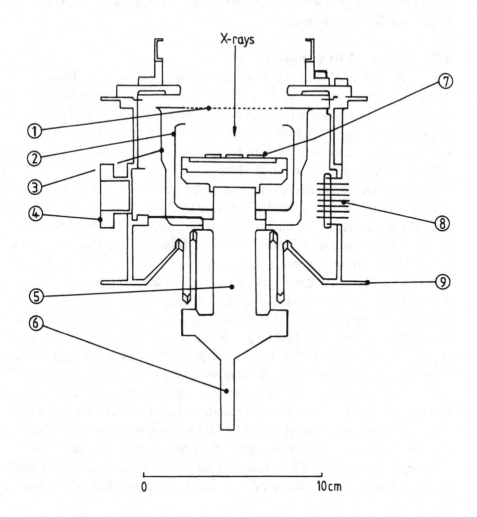

4.6.4   *Applications*

The development of CCDs for X-ray and XUV astronomy proceeds in parallel with work in scientific fields such as optical astronomy, synchotron research (Naday *et al.*, 1987) and high-energy particle physics, where CCD arrays are used as high precision 'vertex' detectors in beam experiments (Bailey *et al.*, 1983; Damerell, 1984; Bocciolini *et al.*, 1985; Bross and Clegg, 1986).

In the last two areas, the detection efficiency (X-ray or particle) of commercial CCDs is at present a constraint so that the deeply depleted CCDs developed by the X-ray astronomers should find ready application in those fields. Similarly, Janesick *et al.* (1985c) have described the benefits of using astronomical X-ray CCDs in high temperature plasma diagnostics. The flash gated CCDs described in section 4.6.2.3 may have applications within the Strategic Defense Initiative (SDI) (Janesick *et al.*, 1986) as well as in solar X-ray astronomy.

4.7   Silicon drift chambers

Silicon drift chambers were first demonstrated at Brookhaven National Laboratory in the USA in 1984 (Gatti and Rehak, 1984; Gatti *et al.*, 1984) and have since received much attention in high-energy particle physics (Rawlings, 1986; Ellison *et al.*, 1987; Kemmer and Lutz, 1987). The first evaluation of these devices for imaging X-ray astronomy has appeared only very recently (Sumner *et al.*, 1988).

Just as a semiconductor diode is the solid state analogue of the gaseous ionisation chamber, so a silicon drift chamber (fig. 4.1d) is the solid state embodiment of its gas-filled namesake. A longitudinal drift field transports electrons in a potential minimum at the midplane of the device to a small readout node. If an appropriate fast 'start' signal is available (necessarily from the hole signal developed on the chamber's surface electrodes if X-rays are to be detected – the so-called self-triggered or self-timing mode – or from an external scintillator if particles are to be counted) the arrival time of the electron signal at the anode is a measure of the x coordinate of the original ionisation. Because its output node has a low capacitance ($\sim 0.1\,\text{pF}$), a cooled silicon drift chamber, like a CCD, may combine good energy resolution (fig. 4.3) with very high spatial resolution ($\Delta x \sim 10\,\mu\text{m}$ fwhm) over areas of several square centimetres. Two-dimensional sensitivity is possible, as are single-sided drift chambers with field electrodes on one surface only. Because the drift chamber electrode structure is less complex than that of a CCD and because the drift chamber is necessarily depleted throughout its $300\,\mu\text{m}$ depth, the X-ray bandwidth of the former device should be much broader. Finally, and very importantly, the time resolution of a silicon drift chamber is restricted only by its signal processing time to the microsecond level, and so should be several orders of magnitude less than the integration time of a CCD.

It is, however, premature to present silicon drift chambers as the future successors to CCDs in astronomical X-ray spectroscopy. Low leakage current chambers are difficult to fabricate. Self-triggered operation at the low signal levels produced by kilovolt X-rays has not yet been demonstrated. Rehak *et al.* (1986) have reported 100 μm rms spatial resolution and $\Delta E = 0.55$ keV in various measurements with 60 keV X-rays.

5

Scintillators, phosphors and NEADs

5.1   Introduction

This chapter describes the uses – past, present and proposed – of three types of solid X-ray converter in soft X-ray astronomy: scintillators and phosphors, which work by the conversion of X-ray energy into visible light and *negative electron affinity detectors* (NEADs), which rely on external photoemission from a surface activated to a state of negative electron affinity. Although the terms scintillator and phosphor are formally synonyms (Thewlis, 1962), we shall adopt the usage prevalent in the detector literature and distinguish between bulk, crystalline materials such as NaI(Tl) and CsI(Na) (scintillators) and thin, granular layers of, for example, the rare earth oxysulphides (phosphors). Phosphors are often identified by a commercial 'P-number'. A partial list of such numbers is given by Gruner *et al.* (1982).

The use of luminescent solids in nuclear physics has a long tradition. Rutherford's nuclear model of the atom (1909), for example, had as its experimental basis the observation, by eye, of α-particle induced light flashes (scintillations) on a zinc sulphide screen. The substitution of a photomultiplier tube for the human observer, which first occurred towards the end of the second world war, produced a sensitive electronic counter for γ-rays and particles, whose operation is described in texts such as those of Curran (1953), Birks (1964) and Knoll (1979).

The first use of a scintillation counter in X-ray astronomy was in a balloon-borne observation of the Crab Nebula in 1964 (Clark, 1965). As described in section 1.2, such balloon payloads were limited, because of atmospheric opacity, to the spectral band $E > 20$ keV, where source fluxes decrease rapidly with increasing X-ray energy. As is made clear in section 5.2.1 – where the principles of scintillation counter operation are described – there are also instrumental factors which limit the use of materials such as NaI(Tl) and CsI(Na) to X-ray energies above about 10 keV. Thus, even when mounted on satellite platforms, the alkali halide scintillators, like the germanium devices described in section 4.4, remain essentially high-energy detectors. Of the other classes of large-area detector described in this text, only xenon-filled gas proportional counters (sections 2.3.1, 2.4.5) compete with them in terms of bandwidth.

Section 5.2.2 describes the construction and operation of a number of balloon- and satellite-borne, imaging and non-imaging, X-ray scintillation counters with appreciable 'soft' response ($E < 50$ keV in the context of this monograph). More extensive reviews of scintillation counters for hard X-ray and γ-ray astronomy are given by Peterson (1975) and Dean (1984).

The many modern applications of inorganic scintillators (especially thallium-activated sodium iodide, NaI(Tl)) in fields other than X-ray astronomy are reviewed by Heath *et al.* (1979). Brooks (1979) surveys the physics and applications of organic scintillators.

No X-ray detector based on a phosphor converter has yet been flown in orbit. Proposals have recently been made, however, for the use in 1–10 keV astronomy of the rare earth oxysulphide phosphors (P43, P44, P45), originally developed for 20–80 keV medical imaging in the early 1970s (Wickersheim *et al.*, 1970; Fishman, 1981). Thin film organic phosphors such as tetraphenyl butadiene (TPB) have also been suggested as the basis of detectors for the soft X-ray and XUV bands. Section 5.3.1 describes the problems of phosphor production and operation. Section 5.3.2 assesses the phosphor-based detector systems (television type and other) currently under development for high resolution imaging X-ray astronomy.

Finally, if phosphors are one of the current instrumental hopes of X-ray astronomy, NEADs represent promise unfulfilled. Section 5.4 describes the development in the mid-1970s of negative electron affinity detectors based on III–V semiconductor photocathodes such as GaAs, which at one time promised to combine high quantum efficiency and spatial resolution with moderate energy resolution.

5.2 Scintillators

The alkali halides NaI and CsI, 'activated' by the introduction of 10^{-2}–10^{-3} mole thallium or sodium impurities, are the scintillators of choice in X-ray astronomy by virtue of their: (i) availability in large formats (> 75 cm diameter; Heath *et al.*, 1979); (ii) relatively high effective atomic number (32 for NaI, 54 for CsI) and hence good X-ray stopping power; (iii) efficient visible light production.

CsI(Na), first introduced in the mid-1960s (Brinckmann, 1965; Menefee *et al.*, 1967), has a light yield comparable to NaI(Tl), first developed in the early 1950s (Heath *et al.*, 1979), but is mechanically more robust, easier to machine and less sensitive to attack by water vapour. NaI(Tl) thus requires encapsulation, whereas CsI(Na) crystals can be stored in low humidity air (Goodman, 1976).

The soft X-ray detection properties of other inorganic scintillators, such as $CaF_2(Eu)$, are described by Aitken (1968a). Newer high-Z materials such as bismuth germanate (BGO: Nestor and Huang, 1975), with light yields $\sim 10\%$ that of NaI(Tl), are presently not available in sufficiently large areas to be considered as the primary component of an astronomical scintillation counter. Plastic scintillators, because of their low fluorescent efficiency to low energy X-rays (Meyerott *et al.*, 1964), are never used for direct photon detection but continue, in association with CsI(Na), NaI(Tl) or BGO crystals, in the role of anti-coincidence counter, a role now outmoded in proportional chamber applications (Cooke *et al.*, 1973), still essential for semiconductor detectors (Chapter 4), and yet to be performed in micro-channel plate work (section 3.4.2).

Table 5.1 contains a list of some scintillator manufacturers.

5.2.1 *Physical principles of scintillator operation*

5.2.1.1 *X-ray interactions*

As in the scintillating gases (sections 2.2.2, 2.5.1.1), light production in the activated alkali halides results from a complex sequence of excitation and de-excitation processes. Only a brief description will be given here.

The role of the impurity in each material is to produce luminescent centres energetically intermediate between the valence and conduction bands of the host crystal. At energies below 100 keV, X-ray interactions with both CsI and NaI are still predominantly by the photoelectric effect (Zombeck, 1982). The photoelectrons produce secondary electron/hole pairs, which, at least in cases where the impurity atom is mercury-like thallium, diffuse to the neighbourhood of the luminescent centres and form

Table 5.1. *Some scintillator manufacturers*

Harshaw Chemical Company, 1945 E 97th Street, Cleveland, Ohio 44106, USA
Nuclear Enterprises, Sighthill, Edinburgh EH11 4BY, Scotland
Hilger Analytical, Westwood, Margate, Kent CT9 4JL, England

long-lived ($\sim 10^{-6}$s) excited Tl$^+$ states which decay with the emission of visible light (Heath *et al.*, 1979). The transfer of energy from the host lattice to the impurity sites is highly efficient. The wavelength of maximum emission is 4200 Å for CsI(Na) (Menefee *et al.*, 1967) and NaI(Tl), and 5700 Å for CsI(Tl) (Bates, 1969). The energy conversion efficiency η (the fraction of the incident X-ray energy which appears as scintillation light) takes the values 0.12 for NaI(Tl), 0.10 for CsI(Na) and 0.05 for CsI(Tl) (Heath *et al.*, 1979). These figures refer to a crystal temperature $\sim 20°C$. The light yields of scintillators are strongly temperature dependent (Heath *et al.*, 1979), which complicates the calibration of balloon-borne instruments which must operate at ambient temperatures of $-40°C$ (Aitken, 1968a) or satellite-borne instruments whose temperature may vary over wide extremes.

The average input energy required to produce one optical photon is given by the expression:

$$w_s = E_s/\eta \tag{5.1}$$

where E_s is the (average) energy of optical emission. This figure of merit may be compared with the parameter w (eq. 2.1) used earlier in the discussions of proportional counter and semiconductor X-ray spectrometers. At room temperature, w_s has values 25, 30 and 44 eV for NaI(Tl), CsI(Na) and CsI(Tl), respectively.

A second important figure of merit is the decay constant of the optical emission, which lies in the microsecond range for most inorganic scintillators and in the nanosecond range for plastic scintillators. By surrounding an alkali halide crystal with a plastic shield and observing the resulting 'phoswich' (phosphor sandwich) with a single photomultiplier tube, one may then use pulse shape discrimination to determine the medium in which energy loss occurred. A phoswich may also be constructed from two alkali halides with dissimilar decay times, e.g. CsI(Tl) (decay time 1.0 μs) and NaI(Tl) (decay constant 0.23 μs) (Peterson, 1975). This background rejection technique is described further in section 5.2.2.1. Bleeker and Overtoom (1979) describe in detail electronic pulse shape analysis for NaI/CsI phoswiches. Not all excited states in a given crystal, however, have the same lifetime. Long-lived phosphorescent states, with decay time constants of minutes or even hours, contribute to the apparent detector background in CsI(Na) telescopes operating below 30 keV (Dean and Dipper, 1981).

Fig. 5.1 shows schematically the fate of the scintillation light in a typical detector. Given:

(i) a high transparency of the crystal to its own optical emission,

(ii) the presence of a reflective coating (which may also act as an environmental shield for the hygroscopic alkali halide), and

(iii) the close matching of the refractive indices of the crystal and its light guide,

many photons will eventually strike the photocathode of a photomultiplier tube (PMT) of gain $\sim 10^6$, resulting in an output pulse whose magnitude is ideally proportional to the input X-ray energy. Multi-alkali optical photo-cathodes are described in detail by Somner (1968) and the operation of conventional multi-dynode PMTs are discussed by Knoll (1979). At the

Fig. 5.1. Non-imaging scintillator operation. Visible light resulting from X-ray interaction in the scintillator crystal passes, either directly or after reflection, through a light guide (e.g. lucite) and strikes the photocathode of a photo-multiplier tube. The energy conversion sequence is the same in imaging detectors incorporating thin phosphor or scintillator layers, where the light guide is either replaced by a fibre optic faceplate or omitted entirely, and the PMT gives way to a microchannel plate or other form of image intensifier.

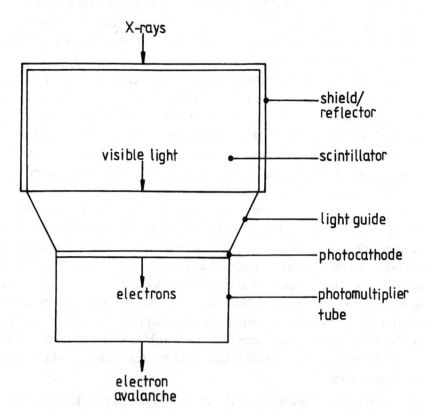

X-ray energies of interest here, scintillation counters are in fact appreciably non-linear. That is, w_s is not independent of E. This effect is described further in section 5.2.1.3.

5.2.1.2 *Quantum detection efficiency*

The quantum detection efficiency (Q in eq. 1.1) of any scintillation counter (or phosphor-based X-ray detector) may be written in the form

$$Q = T_D\{1 - \exp(-\mu_s d)\}\{1 - P(N_p, 0)\} \qquad (5.2)$$

The first term accounts for the absorption of X-rays before entering the active crystal volume. T_D is the composite transmission of the crystal's environmental shield/X-ray window (fig. 5.1) and of any dead layer present on the crystal surface. Since the transmission of thin ($\sim 100\,\mu m$) aluminium or beryllium foils tends rapidly to unity for $E > 10\,keV$ (fig. 2.4) and since deactivated layers, resulting from the exposure of the hygroscopic crystal to humid air, have been found (for CsI(Na)) to have typical thicknesses less than $100\,\mu m$ (Goodman, 1976), T_D is essentially unity for X-ray energies above about 20 keV. These same instrumental factors dictate a low energy cutoff in detector sensitivity between 5 and 10 keV (see figs. 2.4 and 5.2).

The third term in eq. (5.2) is the probability that the X-ray induced light flash produces a measurable PMT output pulse. We denote by $P(N_p, 0)$ the probability that the light flash fails to excite any photoelectron, given that the average number of photoelectrons excited is

$$N_p = (f_s E/w_s)\int_0^\infty p_s(\lambda) q_p(\lambda) d\lambda$$

Here, f_s is the light collection efficiency, which accounts for losses in the coupling between scintillator and photocathode. $p_s(\lambda)d\lambda$ is the probability that the scintillation light has a wavelength in the range λ, $\lambda + d\lambda$ and $q_p(\lambda)$ is the quantum efficiency of the PMT photocathode. The integral term represents the degree of 'spectral matching' between the scintillator's emission spectrum and the photocathode response. The trialkali photocathode S20 (CsNa$_2$KSb: Somner, 1968; Coleman and Boksenberg, 1976), with a peak quantum efficiency of $\sim 25\%$ at 4000 Å is, for example, rather well matched to the emission spectrum of NaI(Tl) and CsI(Na), but even here the value of the integral does not exceed 0.1. Assuming a maximum f_s value of 50% (Peterson, 1975) and adopting the w_s value appropriate to NaI(Tl), we then obtain, with E expressed as usual in kiloelectronvolts

$$(N_p)_{max} \sim 2E \qquad (5.3)$$

Thus, for X-ray energies greater than a few kiloelectronvolts, the term

$(1 - P(N_p,0))$ in eq. (5.2) tends, like the transmission term T_D, to unity. On the other hand, eq. (5.3) tells us that X-rays below ~ 1 keV will be difficult to distinguish from the single electron thermionic noise pulse height distribution of a typical PMT (Aitken *et al.*, 1967; Aitken, 1968a) and so hints at a fundamental low-energy detection limit (see also section 5.2.1.3).

The second term in eq. (5.2) describes absorption in the scintillator itself. Fig. 5.2 compares the mean absorption depth (reciprocal of the scintilla-

Fig. 5.2. Mean absorption depth $1/\mu_s$ as a function of X-ray energy for NaI and CsI, densities, 3.67, 4.51 g/cm³, respectively. The vertical lines indicate the positions of the K absorption edges of iodine (33.17 keV) and caesium (35.98 keV).

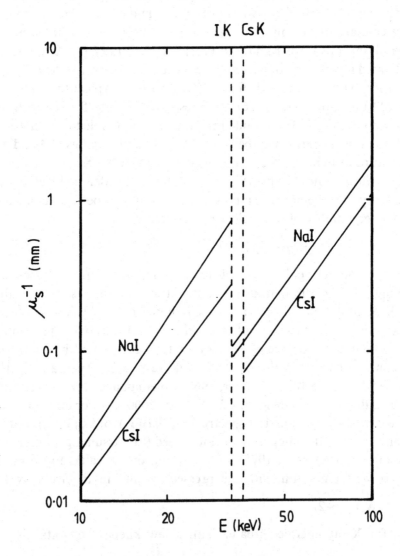

tor's linear absorption coefficient μ_s) in NaI with that in CsI. We see that using a crystal, of either material, of thickness $d=5$ mm would lead to an essentially 100% absorption efficiency up to, and beyond, 100 keV. Even accounting for fluorescence losses above the relevant K edges (fig. 5.2: Carpenter *et al.*, 1976), the detection efficiency of such a crystal would also approach unity over most of the 20–100 keV band, a band in which competing detectors, even xenon proportional counters filled to pressures of several bar, become increasingly transparent with increasing X-ray energy (fig. 2.9). Single crystal thicknesses used in X-ray astronomy have ranged from 1 mm (Clark *et al.*, 1968) to 40 mm (Carpenter *et al.*, 1976).

Both NaI(Tl) and CsI(Na or Tl) scintillators can also be prepared in thin layers (0.5–120 μm thick) by vacuum evaporation (Bauer and Weingart, 1968; Bates, 1969). Such uniform crystalline layers could, in principle, be used for high-resolution imaging of 1–10 keV X-rays in the same way as the granular phosphor layers described in section 5.3. Sams *et al.* (1987) comment that the use of any uniform fluorescent medium allows a relatively high degree of lateral spreading before the light enters the light guide, leading to degraded spatial resolution. The recent development of CsI(Na) scintillation plates with a precise small scale (20 μm) columnar structure (Ito *et al.*, 1987) would appear to overcome this objection to the use of the 'traditional' scintillator materials in imaging devices.

5.2.1.3 *Energy resolution*

The energy resolution of a scintillation counter is determined mainly by photoelectron statistics – the variation in the number of electrons liberated from the PMT photocathode. Assuming that this variation is Poissonian, limits to the fwhm energy resolution can easily be estimated. From eq. (5.3):

$$\Delta E/E = 2.36(1/N_p)^{\frac{1}{2}} \sim 1.67E^{-\frac{1}{2}} \tag{5.4}$$

is the predicted energy resolution of an NaI(Tl) detector with 50% light collection efficiency.

It is immediately obvious that NaI(Tl) (and by extension the other scintillators under consideration here) is a rather poor basis for a non-dispersive X-ray spectrometer, even by comparison with conventional proportional counters (see eq. 2.9). Taking into account the additional problem of non-linear response in the vicinity of the iodine L edges around 6 keV (where w_s changes discontinuously; Aitken *et al.*, 1967), the attractions of NaI(Tl) (or activated CsI) as an X-ray detector are limited to large collecting area (A_s in eq. 1.1) and high quantum efficiency above 20 keV.

The energy resolution of most flight instruments (section 5.2.2) is, in fact, reasonably well approximated by eq. (5.4). For example, Bleeker *et al.* (1967) reported NaI(Tl) resolutions of 50% and 30% at 22 keV and 88 keV, respectively, compared to our predictions of 36% and 18%. Detailed deviations from the $E^{-\frac{1}{2}}$ scaling law have, however, been observed by a number of authors. The CsI(Na) energy resolutions reported by Crannell *et al.* (1977) for the Goddard Space Flight Center detector on the OSO-8 satellite followed a $1.45E^{-0.3}$ law, while Matteson *et al.* (1976), at the University of California, San Diego, measured $1.1E^{-0.38}$ for a balloon-borne NaI(Tl) detector. Recently, Garcia *et al.* (1986) have reported an energy resolution:

$$\Delta E/E = 1.0E^{-\frac{1}{2}}$$

for a NaI(Tl) crystal with a large diameter-to-thickness ratio, directly coupled on to a 40 cm diameter position-sensitive PMT (section 5.2.2). For this crystal geometry, as in the laboratory studies of Aitken *et al.* (1967), the light collection efficiency f_s might be expected to approach unity. Scaling eq. (5.4) by the resulting factor $1/\sqrt{2}$ indeed brings theory and experiment into reasonable agreement.

5.2.2   *Flight instruments*
5.2.1   *Non-imaging instruments*

The design of non-imaging scintillation counters for X-ray astronomy, like that of germanium X-ray detectors (section 4.4), is driven almost entirely by the need to minimise background (B_i in eq. 1.1). Energy resolution and angular resolution are secondary considerations in optimising the geometry of the detector crystal, its collimation and its shielding. The high density and effective atomic number which help make NaI(Tl) and the activated caesium iodides efficient 20–100 keV X-ray detection media inevitably render them sensitive to the high altitude (balloon) and low earth orbit (satellite) radiation environments (section 2.3.2; Peterson, 1975).

The detailed design of low-background scintillation counters has been described by several authors (Kurfess and Johnson, 1975; Peterson, 1975; Matteson *et al.*, 1976; Frontera *et al.*, 1985). Since the pathways by which charged particles or high-energy photons can deposit energies of 20–100 keV in a detector crystal are many and complex, Monte Carlo energy transport programmes (Matteson *et al.*, 1976; Charalambous *et al.*, 1985; Garcia *et al.*, 1986) are commonly used to aid detector design. In every case, there is a tradeoff between (i) passive shielding and (ii) active shielding (active collimation).

The former approach, simply relying on absorption in a shielding mass to cut down the flux of unwanted particles and photons reaching the detector, is a low-cost, high-mass, low-complexity option. It carries the risk

Fig. 5.3. Background rejection by active collimation and phoswich pulse shaping (after Anderson, 1972; Peterson *et al.*, 1972). Crystals A and C constitute a phoswich, viewed by a single PMT, and are optically isolated from shield crystal B. Energy may be deposited in the detector crystal A (usually NaI(Tl)) by a variety of routes: (i) by direct entry of source or diffuse background photons through the (CsI(Na)) collimator, which is optically part of the shield. The design of drilled scintillator collimators is discussed by Aitken (1968b). (ii) By cosmic ray ionisation loss. Prompt particle events may be eliminated by using signals from crystals B and C (second component of the phoswich) in anti-coincidence with the detector crystal. (iii) By rear or side entry of a photon after Compton scattering in either of crystals B or C. Such events can again be distinguished by anti-coincidence or pulse shaping techniques. Pathways (iv) represent passive photon absorption in the shield. Pathway (v) represents leakage through the shield. Events (vi) correspond to Compton scattering in the detector crystal followed by photoelectric absorption in crystal C. The electronic logic for such detectors is described by Anderson (1972), Peterson (1975) and Matteson *et al.* (1976).

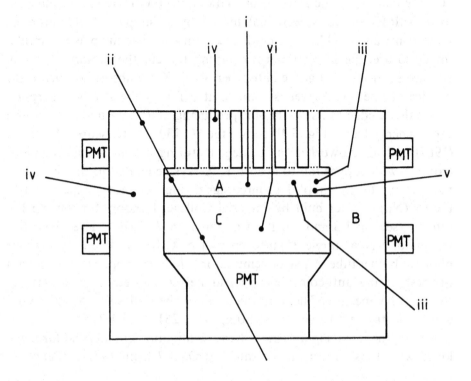

that adding material in the vicinity of the detector may increase the local production of secondary particles or photons. The latter approach, using a scintillating medium for the shielding mass, has a number of advantages regarding the range of event types which may be vetoed (Carpenter *et al.*, 1976). It is a high-cost, low-mass, high-complexity option, first employed by Frost *et al.* (1966), favoured in most recent scintillator designs and illustrated schematically in fig. 5.3.

A simple criterion for the optimum thickness of shielding is that the leakage of high-energy photons through the shield should equal the diffuse flux entering the counter aperture. Then (Peterson, 1975)

$$\Omega/4\pi > \exp(-\mu_{sh}(E)t)$$

where Ω is the aperture size in steradians (eq. 1.1) and μ_{sh} is the energy-dependent linear absorption coefficient of the shield material. Applying this criterion at 100 keV to a typical detector with 6° fwhm collimation (Matteson *et al.*, 1976; Crannell *et al.*, 1977), and assuming that the (active) shield material is CsI (fig. 5.3), we find that the necessary thickness is $t \sim 1$ cm.

Figs. 5.4 and 5.5 illustrate some of the simpler detector configurations which were flown in the 1960s and early 1970s. Fig. 5.4a shows the 20–100 keV balloon-borne detector of Clark *et al.* (1968). Here, four independent NaI(Tl) modules were collimated by a simple 12° fwhm brass honeycomb and shielded by a graded lead/tin absorber and plastic scintillator enclosure, the last component serving to veto the prompt effects of charged particles. In a graded absorber, weak X-ray absorption below the K edge of one metal layer is compensated for by the strong absorption above the K edge of its neighbour, producing uniform attenuation over a broad energy band. Fig. 5.4b shows the 20–130 keV telescope of Bleeker (1967) and his co-workers at Leiden, featuring a conical tin collimator (covered with copper foil to absorb K X-rays generated in the tin) and a rotating shutter to 'chop' the source signal. Fig. 5.5 shows Experiment F, the CsI(Na) counter built by Imperial College, London, for the Ariel V satellite (section 1.3 and fig. 1.5; Carpenter *et al.*, 1976). Here, the active collimator was a simple 35 cm deep well of CsI(Na) viewed by a single photomultiplier tube. A plastic scintillator 'window' in optical contact with the rest of the anti-coincidence shield helped to veto charge particles entering the aperture. The 4 cm thickness of the CsI(Na) detector crystal gave an instrument response extending from 26 keV to 1.2 MeV.

Among the most sophisticated X-ray scintillation counters so far developed were those flown on the satellites OSO-7 from 1971–3 (Peterson,

Fig. 5.4. (a) Scintillation counter used for 20–100 keV observations of the constellation Cygnus (Clark *et al.*, 1968) from a high altitude balloon. (*Courtesy G.
Clark.*) (b) University of Leiden scintillation counter with rotating source-
occulting shutter (Bleeker *et al.*, 1967). (*Courtesy J. Bleeker.*)

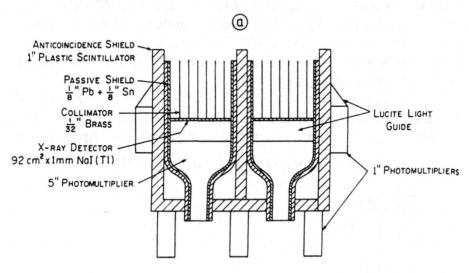

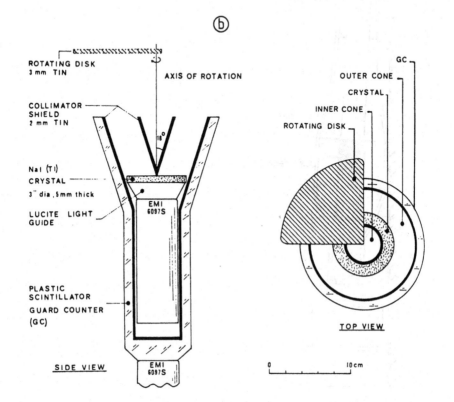

1975; Rothschild *et al.*, 1981), OSO-8 from 1975–8 (Crannell *et al.*, 1977) and HEAO-1 (Experiment A4: Peterson, 1975; Primini *et al.*, 1981). All three devices had essentially the geometry illustrated in fig. 5.3. The OSO-7 detector consisted of a large-area (64 cm²), 1 cm thick NaI(Tl) crystal actively shielded by CsI(Na). The OSO-8 scintillation spectrometer consisted of two separate 27.5 cm² CsI(Na) crystals, one equipped with a drilled CsI(Na) collimator (fig. 5.3: Aitken, 1968b) and the other, a completely shielded background monitor.

Figs. 5.6a and b compare the OSO-7 and HEAO-1 A4 observations of the Perseus Cluster, both of which show an excess of hard X-ray emission in addition to the thermal Bremmstrahlung spectrum (Appendix B)

Fig. 5.5. Ariel V Experiment F (Carpenter *et al.*, 1976). (*Courtesy J. Quenby.*)

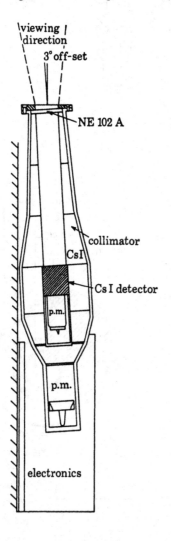

Fig. 5.6. Scintillation counter observations. (a) OSO-7 8-432 keV spectrum of the Perseus Cluster (Rothschild *et al.*, 1981). (*Courtesy R. E. Rothschild.*) (b) HEAO-1 Experiment A4 spectrum of the Perseus Cluster (Primini *et al.*, 1981), compared with previous lower energy measurements. kT in each case is the characteristic energy of the best fit thermal Bremsstrahlung spectrum (Appendix B) at energies below 20 keV (*Courtesy R. E. Rothschild.*) (c) Hard X-ray spectrum of Her X-1 (Trümper *et al.*, 1978). (*Courtesy J. Trümper.*)

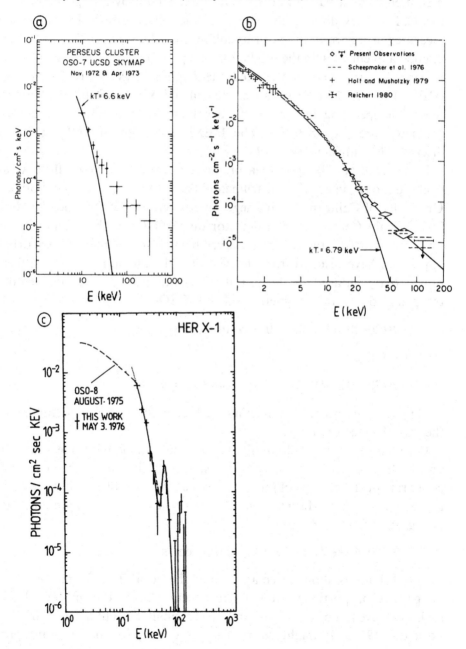

observed at lower energies. Such featureless spectra are typical of most scintillation counter observations. Fig. 5.6c shows the relatively rare positive detection of a line feature using a scintillation counter (Trümper *et al.*, 1978). The 58 keV feature in the spectrum of Hercules X-1, attributed to electron cyclotron emission in the plasma of a rotating neutron star, was measured using a balloon-borne NaI counter built by groups at the Max Planck Institut, Garching, and the Astronomisches Institut, Tübingen. The measured energy resolution of this collimated detector was 22% at 60 keV, in good agreement with the prediction of eq. (5.4). More recently, the same research groups have constructed the High Energy X-ray Experiment (HEXE) for the Mir/Kvant space station. HEXE is a four-fold NaI/CsI phoswich operating in the 15–250 keV band, with $\theta_{\frac{1}{2}} = 1.6°$ and a total geometric area $A_s = 800 \, \text{cm}^2$. The HEXE spectrum of SNR 1987A is discussed by Sunyaev *et al.* (1987).

Kinzer *et al.* (1978) report measurements of the 20–165 keV diffuse X-ray background (B_d in eq. 1.1). Instrumental background levels (B_i) in both the University of California, San Diego, instrument on OSO-7 and Goddard Space Flight Center (GSFC) detector on OSO-8 have been analysed by Dyer *et al.* (1980). Much of the observed background in their 500 km orbits appears to have resulted from delayed particle interactions; that is, from radioactivity induced in the scintillator materials by energetic protons trapped in the earth's magnetic field. For OSO-7:

$$B_i(20\text{–}40 \, \text{keV}) = 2 \times 10^{-2} \, \text{counts/cm}^2 \, \text{s}$$

while for OSO-8:

$$B_i(20\text{–}40 \, \text{keV}) = 1.5 \times 10^{-1} \, \text{counts/cm}^2 \, \text{s}$$

the higher figure probably reflecting the higher effective atomic number of the OSO-8 detector crystal.

For relatively short balloon flights, induced radioactivity is less important. In fact, at balloon altitudes, no single effect dominates the overall count rate in an actively collimated phoswich detector of the kind shown in fig. 5.3, for which Matteson *et al.* (1976) have estimated an ultimate background limit of $8 \times 10^{-5} \, \text{counts/cm}^2 \, \text{s keV}$ or:

$$B_i(20\text{–}40 \, \text{keV}) = 1.6 \times 10^{-3} \, \text{counts/cm}^2 \, \text{s}$$

This last figure is about twice as good as the in-orbit performance of the Ginga LAC proportional counter array (section 2.3.1). It is only ever likely to be achieved, however, with relatively small detectors ($A_s < 100 \, \text{cm}^2$; Peterson, 1975). It might seem that large-area, low-background proportional counters, filled to rather higher pressures (>5 bar) than those currently used in the 2–30 keV band, could, in the future, completely supplant scintillation counters for all low-resolution spectral studies below 100 keV, despite their low quantum efficiencies (~20% at 100 keV). In practice, the competition will surely be between imaging devices; between high-pressure imaging proportional counters of the type proposed by Ubertini *et al.* (1983) and the scintillation counters described in the next section.

5.2.2.2 *Imaging instruments*

Two types of imaging scintillation counter are currently being developed for use with coded mask telescopes (section 1.5).

In the first type, the light signals (amplitudes P_A, P_B) observed by photomultiplier tubes coupled to the ends of a long NaI(Tl) bar, whose surface has been specially treated, are used to determine the positions of photon interactions. In such a 'light dividing' encoder (cf. the charge division encoders of section 2.4.1.1), assuming exponential attenuation of the scintillation light along the bar

$$x = (1/2\alpha) \ln (P_B/P_A)$$

where α is the light attenuation per unit length and x is the interaction position relative to the centre of the bar (Carter *et al.*, 1982; Charalambous *et al.*, 1984). Stacking a number of such bars in parallel then produces a detection plane sensitive in two dimensions, with spatial resolution in one dimension set by the bar width (~5 cm). Measurements in the other dimension, along the bar length (~50 cm), indicate

$$\Delta x = 41 \ E^{-\frac{1}{2}} \text{cm fwhm}$$

which makes the technique rather unattractive for the energy band of principal interest here. The 'striped' nature of such NaI(Tl) detection planes has inspired the name ZEBRA for the 100 keV–10 MeV Anglo–Italian balloon experiment in which they are to be used. The ZEBRA payload also includes a complementary high-pressure proportional counter (Ubertini *et al.* 1983).

Researchers at the Harvard-Smithsonian Center for Astrophysics (Garcia *et al.*, 1986; Grindlay *et al.*, 1986) have recently described a second approach to imaging with NaI(Tl), directed specifically to the 20–100 keV band. In their balloon-borne EXITE (Energetic X-ray Imaging Telescope Experiment) detector, a thin (0.64 cm), large diameter (34 cm) NaI(Tl) crystal is bonded directly to the convex front surface of a two-stage image intensifier tube (Rougeot *et al.* 1979), originally developed for medical imaging. After photon-to-electron conversion and electrostatic demagnification in the first stage of the intensifier, the image is read out from the second stage by a silicon diode charge divider (Laegsgaard, 1979) whose electrode arrangement is identical to that of the resistive anode illustrated in fig. 2.20b. Measured spatial resolution follows the law (Garcia *et al.*, 1986):

$$\Delta x = 6.6\ E^{-\frac{1}{2}}\,\text{cm fwhm}$$

much lower than the light division technique described above. The EXITE crystal geometry also lends itself to good energy resolution as has already been discussed (section 5.2.1.3). Given an estimated background level:

$$B_i(20\text{--}40\,\text{keV}) = 1.3 \times 10^{-2}\ \text{counts/cm}^2\,\text{s}$$

an effective area (QA_s) of up to $450\,\text{cm}^2$ and the sensitivity advantage associated with imaging at an angular resolution of $0.3°$, results from EXITE should easily match those obtained from the satellite-borne, non-imaging scintillation counters of the 1970s.

5.3   Phosphors

The physics of these X-ray converters is essentially the same as that of the scintillators described in previous sections. Moreover, the key functional elements of any phosphor-based X-ray detector are the same as those of the schematic scintillation counter illustrated in fig. 5.1. It is in their astronomical applications that the two classes of material – phosphors and scintillators – differ enormously. While large single crystals of NaI(Tl) and the activated caesium iodides are valued for their stopping power in hard X-ray and γ-ray astronomy, work with thin (largely polycrystalline) layers of: (i) the rare earth phosphors, notably P43 ($Gd_2O_2S(Tb)$), P44 ($La_2O_2(Tb)$) and P45 ($Y_2O_2S(Tb)$), each with terbium activation at the $\sim 10^{-3}$ mole level, and (ii) organic phosphors such as TPB (1,1,4,4 tetraphenyl–1,3 butadiene, $C_{28}H_{22}$) has been directed towards high resolution, soft X-ray imaging. The rare earth phosphors listed above are used commercially in cathodoluminescent screens for cathode ray tubes, in medical X-ray imaging (Fishman, 1981) and in synchotron research (Arndt, 1982). X-ray detectors based on these materials have, potentially, the highest spatial resolution of all photon counting X-ray imagers, limited to a few times the phosphor grain size ($>1\,\mu\text{m}$) in films a few grain layers thick. Research using uniform thin films of TPB and similar fluorescent materials (e.g. sodium salicylate ($NaC_7H_5O_3$)) is somewhat less advanced

(Bedford and St. J. Mann, 1984) and likely to produce detectors of poorer resolution, because of the problem of unconstrained lateral light spreading discussed in section 5.2.1.2 (Arndt, 1982; Sams *et al.*, 1987). TPB and sodium salicylate are commonly used as detectors for ultraviolet radiation (Samson, 1967; Burton and Powell, 1973).

5.3.1 *Physical principles of phosphor detector operation*

Luminescence mechanisms in some rare earth phosphors are described in detail by Buchanan and Wickersheim (1968) and by Wicker-sheim *et al.* (1970). Decay constants vary enormously with phosphor composition (Chappell and Murray, 1984). For P43, the decay constant is a few hundred microseconds (Arndt, 1982), which would limit the count rate to the $\sim 10^3$/s level in detectors incorporating that phosphor (Schwarz *et al.*, 1985a). X-ray energy conversion efficiencies in the best 'P' phosphors are comparable with that of NaI(Tl) – 15% for P43 (Arndt, 1982), 7–9% for P44 (Kalata, 1982). Wavelengths of peak phosphor emission fall around 5500 Å, so that w_s, the average input energy required to produce one optical photon (eq. 5.1), lies in the 16–33 eV range for these materials.

Methods of phosphor preparation are described by Chappell and Murray (1984) and by Sams *et al.* (1987). The commonest method of producing a fine-grain phosphor screen is by sedimentation, whereby grains are suspended in an appropriate binder (e.g. potassium silicate) – solvent (e.g. acetone, water) solution and allowed to settle onto the desired substrate, usually a fibre optic faceplate. Once the solvent has evaporated, a pinhole-free, densely packed phosphor screen should remain. Other techniques include brushing the phosphor grains onto a uniform epoxy layer with a fine haired paintbrush (Sams *et al.*, 1987). TPB is prepared by vacuum evaporation and sodium salicylate by spraying a saturated solution in methanol (Bedford and St. J. Mann, 1984).

Typical screen 'thicknesses' (density times linear thickness) for the 1–10 keV band lie in the range 0.1–20 mg/cm². Since visible light attenuation is least in thin phosphor layers and X-ray attenuation is greatest in thick layers, there is generally an optimum thickness, corresponding to the maximum quantum efficiency, for each X-ray energy and each material (Chappell and Murray, 1984). Fig. 5.7 compares the X-ray absorption efficiency of 10 mg/cm² of P43 with that of equivalent layers of CsI and ZnS (which remains a useful phosphor material more than eight decades after its original use). At 6 keV, the X-ray absorption efficiency of P43 exceeds 85%, as recently confirmed by Schwarz *et al.* (1985a).

The greatest number of photoelectrons (eq. 5.3) so far reported with any phosphor matched to an appropriate PMT is:

$$(N_p)_{max} = (2.3-4.0)E$$

or 14–24 photoelectrons per 6 keV X-ray for 5.8 mg/cm² of P45 (Chappell and Murray, 1984; Kalata *et al.*, 1985). The lower figure refers to an uncoated phosphor; the higher, to a laquer coated layer with an aluminised polypropylene reflector, for maximum light collection (fig. 5.1, section 5.2.1.2). Such figures indicate that the energy resolution of phosphor-based X-ray detectors should be at the same relatively crude level as NaI(Tl). No measurements of energy resolution have yet been reported, beyond an indication that the response of P44 and P45 is approximately linear with

Fig. 5.7. Comparison of the X-ray absorption efficiency of the phosphors Gd₂O₂S(Tb) (P43), ZnS (P11, if activated by silver; P2, if activated by copper) with that of CsI (Arndt, 1982). All layers characterised by $\rho_s d = 10$ mg/cm², where ρ_s is the density (7 g/cm³ for P43) and d is the linear thickness of the layer. μ'_s is the mass absorption coefficient of the phosphor. (*Courtesy U. Arndt.*)

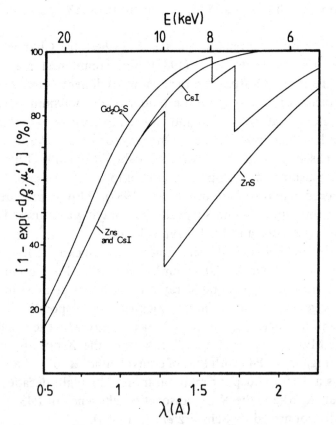

energy E (Chappell and Murray, 1984; Kalata *et al.*, 1985). The number of visible photons produced per X-ray incident to TPB decreases with X-ray energy in the range 0.28–8 keV because of the low stopping power of such hydrocarbon layers to kilovolt X-rays (Bedford and St. J. Mann, 1984).

5.3.2 *Detector concepts*

The most highly developed phosphor-based X-ray detector for astronomy is currently the Harvard-Smithsonian Center for Astrophysics PXI (Phosphor X-ray Imager: Kalata, 1982; Chappell and Murray, 1984; Kalata *et al.*, 1985). PXI is one of a family of television-type detectors which have been developed for X-ray, visible light, UV and particle detection. Its first use in X-ray astronomy will be in conjunction with a normal-incidence multilayer telescope (section 1.5), optimised for a narrow wavelength range centred on 63 Å, and flown on a sounding rocket to image active regions of the sun (Golub *et al.*, 1985). The PXI is illustrated

Fig. 5.8. Phosphor-based X-ray imaging detectors. (a) PXI (Kalata, 1982; Chappell and Murray, 1984; Kalata *et al.*, 1985). In this version of the detector, the diameter of the phosphor is 13 mm. (*Courtesy K. Kalata.*) (b) Phosphor detector with microchannel plate readout (Schwarz *et al.*, 1985a). WSA = wedge-and-strip anode readout element (fig. 2.22c). The detector is shown illuminated through a pinhole test mask. (*Courtesy H. Schwarz.*)

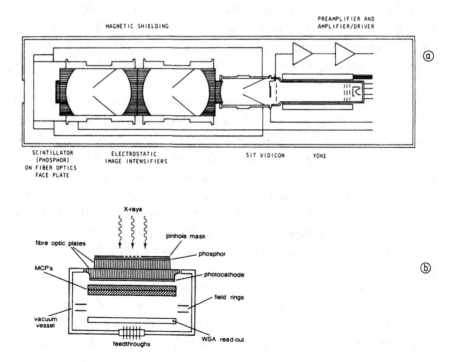

schematically in fig. 5.8a. Light is produced in a phosphor layer deposited on the planar entry side of the first fibre optic faceplate in an image intensifier chain. This light excites photoelectrons from a transmission photocathode deposited on the concave exit surface of the fibre optic. These electrons are then accelerated onto the phosphor screen of an inverting image intensifier module. The electron image leaving a second such module is finally read out by a SIT (silicon intensifier target: Coleman and Boksenberg, 1976) vidicon. Very high spatial resolution ($\sim 3 \, \mu m$ fwhm: Sams *et al.*, 1987) can be achieved by using a 1:4 expanding fibre optic faceplate at the site of the X-ray conversion to match the intrinsic phosphor resolution to that of the image intensifiers and SIT vidicon.

Whereas the other photon-counting detectors described in this monograph feature one, two or, at most, three energy conversions, in the PXI detector there are six such stages. The electronics required to 'drive' the PXI detector reflect in their complexity the physics of the multistage approach to X-ray imaging: these electronics are described in detail by Kalata (1982) and by Kalata *et al.* (1985). Since the detector output is in fact a video image, telemetry requirements for a space-borne PXI would appear to be at least as demanding as for the CCD cameras described in section 4.6. Like a CCD detector, PXI can be operated both in a single photon location mode and in an integrating ('analogue-to-digital') mode, whereby the intensity in each pixel is digitised at appropriate intervals. Such flexibility allows a huge range of event rates – $1–10^{12}$ counts/s – to be covered by one programmable detector (Kalata, 1982). Of course, seven decades of that response are unlikely ever to be necessary in cosmic X-ray astronomy.

A simpler, more compact phosphor-based detector has recently been proposed in outline by researchers at MSSL, for possible application to ESA's XMM mission (Schwarz *et al.*, 1985a). Here, the phosphor converter is a layer of P43 and the image intensifier, a position-sensitive microchannel plate detector (fig. 5.8b). Spatial resolution in a prototype detector ($\sim 160 \, \mu m$ fwhm) was dominated by the relatively crude proximity focusing of photoelectrons leaving the optical photocathode. Ultimately, such a detector would be limited in resolution by the pore size of the MCPs to a level $\sim 15 \, \mu m$, still much higher than the intrinsic phosphor limit accessible in the PXI design.

Finally, we should note that phosphor overcoating may provide a relatively low-cost method of extending the response of front-illuminated CCDs (Germer and Meyer-Ilse, 1986) and other silicon devices (Zutavern *et al.*, 1980) below 1 keV in X-ray energy. The visible light emitted from a phosphor layer coated onto a CCD or photodiode array can penetrate the device electrode structure which acts as a barrier to sub-kilovolt X-rays (section 4.6.2.3). The red light from P50 (Y_2O_3(Eu)) is particularly well-matched to the optical response of silicon sensors (Zutavern *et al.*, 1980; Gruner *et al.*, 1982). Germer and Meyer-Ilse (1986) have reported a creditable 9% quantum efficiency at 0.28 keV for a P43-coated CCD operating in integrating mode as a detector for X-ray microscopy. For some 0.1–1.0 keV applications the 'fix' of phosphor coating a standard chip may prove more cost effective than the technologically demanding development of back-illuminated CCDs currently being pursued in astronomy.

5.4 NEADs

Any survey of the instrumentation literature of the late 1970s, particularly those papers 'comparing and contrasting' detectors for future satellite missions, will inevitably throw up references to NEADs – negative electron affinity detectors – devices with the following highly desirable properties: (i) 15 μm spatial resolution over 25 mm diameter fields, (ii) high 0.1–7 keV quantum efficiency, (iii) moderate energy resolution ($\Delta E/E = 20\%$ at 1 keV; 100% at 6 keV).

Present-day detector reviews, however, rarely even mention NEADs: a brief description of their 'rise and fall' may provide a healthy reminder that there is no inevitability regarding the eventual success of any of the newer, untried detector concepts discussed in this monograph.

The study of NEADs for imaging X-ray astronomy began at the Harvard-Smithsonian Center for Astrophysics in the early 1970s. The goal was to produce a high-resolution detector for the HEAO-B (Einstein) mission with X-ray quantum efficiency much higher than that (then) obtainable with microchannel plates.

Semiconducting compounds composed of elements from the third and fifth columns of the periodic table (e.g. GaAs) can be activated to a state of negative electron affinity (NEA) by treatment of their surfaces with caesium and oxygen. That is, the top of the conduction band in the bulk can be made to lie above the vacuum level. Conduction band electrons can therefore escape from the surface when an X-ray is absorbed, even if the photon interaction takes place deep in the material (Van Speybroeck *et al.*, 1974). Electron diffusion lengths in GaAs and GaAsP are of order 2 μm, compared with 0.02 μm in conventional X-ray photocathodes such as CsI (section 3.3.3). The energy required to produce one internal free electron, *w* (eq. 2.1), is 4.2 eV in gallium arsenide (Cuzin, 1987). GaAs is, of course, one of the class of room temperature semiconductor materials (Eberhardt *et al.*,

1971) discussed in section 4.5. The quantum detection efficiency of NEA materials at photon energies of a few kiloelectronvolts is thus close to unity. Moderate energy resolution follows from the fact that the most probable number of emitted photoelectrons increases monotonically with E, just as it does, less steeply, for CsI (section 3.3.4). The apparently anomalous degradation of energy resolution with increasing E results from the preponderance of 'partial' events (in which only a few of the excited electrons manage to escape) when the mean photon absorption depth is large. The same effect is observed using CsI photocathodes. Cooling an NEA photocathode to $-30°C$ (Bardas et al., 1978) efficiently suppresses the detector background arising from thermionic emission.

Fig. 5.9 shows the NEAD detector concept proposed by Van Speybroeck et al. (1974). The key element is the GaAs photocathode operated in transmission mode (i.e. with X-rays focused by a grazing-incidence telescope on the front surface producing an electron image at the rear 'activated' surface). A further image intensifier stage would of course be necessary to record the visible light signal from the phosphor screen.

Fig. 5.9. NEAD X-ray detector (Van Speybroeck et al., 1974). The electron image leaving the activated surface is accelerated through several kilovolts and strikes a phosphor screen. The volume between GaAs wafer and phosphor must be highly evacuated (pressure $\sim 10^{-10}$ Torr). Before launch, the volume to the left would also be evacuated (to a pressure $\sim 10^{-5}$ Torr), with the vacuum door shut, to avoid mechanical stress on the photocathode. Once in orbit, the door would be opened. (*Courtesy L. Van Speybroeck. © 1974 IEEE.*)

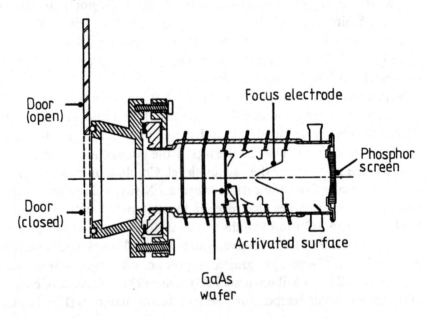

Although the interaction of X-rays with NEA materials could be thoroughly quantified in terms of the model developed by Van Speybroeck *et al.* (1974) and extended by Bardas *et al.* (1978), no such position-sensitive NEAD was ever demonstrated by the Center for Astrophysics researchers. The stumbling block appears to have been the need to maintain the NEA surface in an extremely high vacuum ($\sim 10^{-10}$ Torr) in order to prevent contamination which might 'kill' the negative electron affinity property. This constraint would have been much easier to meet in the sealed, mica-window PMTs used for the initial laboratory investigations of NEAD properties than in any practical, large-area imaging device. Timothy (1988) has speculated that NEA photocathodes may be compatible with future generations of microchannel plate, manufactured from glasses bakeable to very high temperatures ($>300°C$).

In any case, NEADs could not be made to work within the cost and schedule constraints of the HEAO-B programme (Tucker and Giacconi, 1985) and have never yet been revived. In the post-Einstein era, other detectors with much less severe environmental requirements – CCDs – have effectively filled their niche in the X-ray astronomer's instrumental armoury.

6

Single photon calorimeters

6.1 Introduction

Probably the most radical advance in X-ray instrumentation in the past five years has been the development of the single photon calorimeter, in which X-rays are detected via the temperature pulses they induce in a small ($<1\,mm^3$) absorber, cooled to a fraction of a degree kelvin.

The detection of individual 5.9 keV X-rays (fig. 6.1) was first demonstrated by groups at NASA's Goddard Space Flight Center (GSFC) and the University of Wisconsin in 1984, using a silicon microcalorimeter operating at 0.3 K (McCammon *et al.*, 1984). This work was specifically directed towards the production of a high-efficiency, non-dispersive focal plane spectrometer with energy resolution comparable to that of a Bragg crystal. It can, however, still be seen as the culmination of several decades' research in fields other than X-ray astronomy, originally in nuclear physics and latterly in infrared astronomy. Andersen (1986) and Coron *et al.* (1985a) trace calorimeter development back as far as 1903, and the radioactivity studies of Pierre Curie. They record how, by the mid-1970s, the sensitivity (in detectable watts) of IR bolometers operating at liquid helium temperatures, where heat capacities are very low, had reached the point where Niinikoski and Udo (1974) could identify the extraneous spikes seen in the output of balloon-borne bolometers with local heating produced by the passage of cosmic rays. Niinikoski and Udo appear to have been the first to suggest that it might be possible to thermally detect single photons or particles, rather than continuous fluxes.

In this chapter, we shall outline the physical principles of calorimeter operation (section 6.2) and describe the problems involved in building a satellite-borne thermal X-ray detector (section 6.3), referring in both sections to results from the GSFC/Wisconsin AXAF detector development programme. Finally, section 6.4 reviews the recent upsurge of interest in single photon, single particle calorimeters for applications other than X-ray astronomy.

6.2   Calorimeter operation
6.2.1   *Energy resolution: expected and achieved*
Single photon calorimeters work by the low-noise conversion of absorbed energy to heat. Such devices consist essentially of an absorber (linked to a heat bath at ~ 0.1 K) and one or more thermistors, each linked, through a load resistor, to a low-noise amplifier with a cooled (80 K) first stage. The temperature rise induced in the former element by the absorption of an individual photon or particle produces voltage waveforms at the outputs of the latter from which energy information may be abstracted. Figs. 6.2a–c show schematically some X-ray calorimeter designs proposed

Fig. 6.1. Voltage pulse resulting from the absorption of a single 5.9 keV photon in the prototype silicon calorimeter of McCammon *et al.* (1984). (*Courtesy D. McCammon.*)

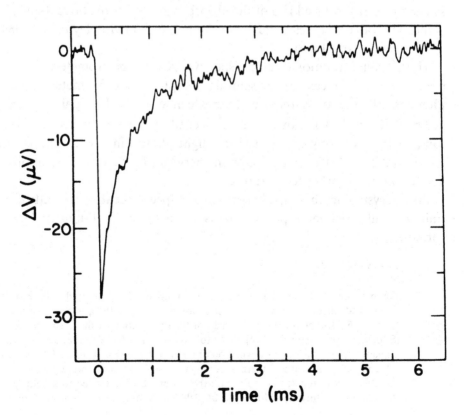

to date, together with one-dimensional models incorporating their essential features.

The 'flow' of absorbed energy between competing physical processes in the absorber material may be rather complex (Andersen, 1986). Loss mechanisms such as photoemission, fluorescence and charge trapping onto crystal defect sites will inevitably account for some fraction of the energy. The rest (hopefully, from the point of view of calorimeter operation, nearly all) will end up as heat. Then, providing that amplifier noise and Johnson noise in the load resistor(s) can be made negligibly small, the energy resolution of the thermal spectrometer is determined by thermodynamic energy fluctuations (fluctuations in phonon number) within the absorber and the links connecting it to its heat bath. This limiting resolution is given by the expression (Moseley *et al.*, 1984):

$$\Delta E = 2.36 \xi (k T_0^2 C)^{\frac{1}{2}} \tag{6.1}$$

where C is the heat capacity (in Joules per kelvin) of the detector at a heat sink temperature T_0, and k is Boltzmann's constant. ξ is a detector constant, dependent largely on the properties of the thermistor which senses the X-ray induced rise of the absorber temperature above its steady value. The value of ξ lies typically in the range 1–3 (Moseley *et al.*, 1984; McCammon *et al.*, 1986).

Thus, while traditional non-dispersive X-ray spectrometers 'count' electrons or, in the case of the scintillators, photons, calorimeters 'count' phonons. The energy, w, required to create an electron–hole pair in silicon is 3.6 eV (Table 4.1) and over 20 eV is needed to create either an electron–ion pair in a counting gas or a visible light photon in a solid scintillator (sections 2.2.1, 5.2.1.1). Mean phonon energies $k T_0$, however, are of order 10^{-4} eV at an absorber temperature of 1 K.

For a crystalline absorber, the dominant specific heat of the lattice (in units of joules per mole per kelvin) is given by the well-known Debye formula:

$$c = 1944 (T_0/T_D)^3 \tag{6.2}$$

Fig. 6.2. Single photon calorimeters and 1-d models. (a) Proposed high resolution etched silicon calorimeter (McCammon *et al.*, 1984, 1986; Moseley *et al.*, 1984, 1985; Holt *et al.*, 1985). The legs of the device connect the absorber mass (Si possibly overcoated with 2 μm Bi) to a 0.1 K heat sink. External leads to the implanted thermistor not shown. Looking along the diagonals *ac* or *bd* suggests a 1-d model with identical thermal conductances G_{link} at each end and a central, pointlike thermistor. (b) Variation on the design of (a), intended to minimise thermal non-uniformity (Moseley *et al.*, 1984). The outputs of the four identical

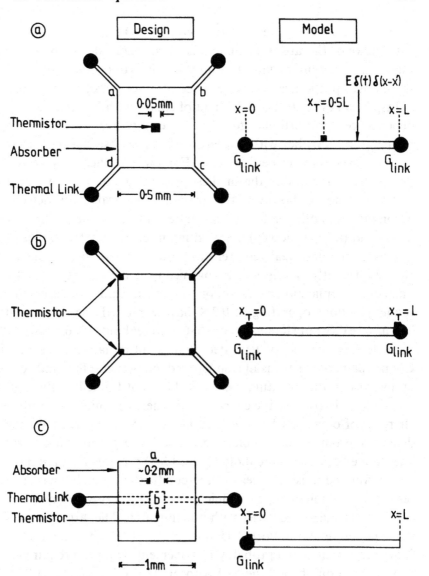

caption for fig. 6.2 (*cont.*)

vertex thermistors may be summed or otherwise manipulated in order to obtain a position independent energy signal. The corresponding 1-d model has thermistors at the line-ends $x/L = 0$, 1. (c) Composite diamond–germanium bolometer (Coron *et al.*, 1985a, b; Stroke *et al.*, 1986). The diamond absorber is supported from below by a doped Ge thermistor connected to a heat sink by two legs. The corresponding 1-d model, arising from an examination of the central sections *ab* or *bc*, has a coincident thermistor and thermal link at one end and no heat loss from the other end.

In each of the models, developed for the study of the effects of thermal non-uniformity (Fraser, 1987), energy is deposited as a temporal delta function at coordinate $x = x'$.

where T_D is the Debye temperature, equal to 640 K for silicon and 2220 K for diamond, the absorber materials mostly considered for microcalorimeter construction to date. Fiorini and Niinikoski (1984) have tabulated the thermal properties of a wide range of other possible materials.

Inspection of eq. (6.1), in the light of eq. (6.2), thus reveals a very strong dependence (proportional to $T_0^{5/2}$) of energy resolution on operating temperature, together with a need to minimise absorber mass in order to optimise spectrometer performance. For some applications, there may be no serious constraint on the smallest useful detector volume. For imaging X-ray astronomy, however, the area of any focal plane detector must accommodate both the finite angular resolution of the grazing-incidence optics and (more seriously) the pointing jitter of its satellite platform.

This 'minimum area' requirement would appear to rule out a second class of recently developed, cryogenically cooled X-ray detector from immediate application in astronomy. Tin/tin oxide/tin superconducting tunnel junctions operating at 0.3 K have exhibited energy resolutions as low as 67 eV fwhm (Kraus *et al.*, 1986; Twerenbold, 1986, 1987, 1988). In these devices, the energy deposited by incident radiation is used to 'break' Cooper pairs of electrons in a superconducting metal film and so produce an increase in the quantum mechanical current tunnelling through a thin (20 Å) oxide barrier. w, the characteristic energy required to produce a free charge is of order 0.001 eV, so that, like the calorimeter, superconducting detectors promise to combine very good energy resolution with high quantum efficiency. Twerenbold (1987, 1988) presents an analysis of signal generation and noise sources in superconducting tunnel junctions. Energy resolution is limited by electronic noise which arises from the junction capacitance and hence varies with junction area. The largest X-ray detecting junction so far reported (Kraus *et al.*, 1986) had an area of only 0.088 mm^2. An array of probably 1000 elements would be required to cover a useful fraction of (say) the XMM focal plane (Twerenbold, 1987). High-temperature (90 K) superconducting materials offer no great advantage over tin as a detection medium, since the junction must be operated at a temperature which is only a small fraction of its transition temperature, and since the value of w in high-temperature superconductors is likely to be much higher than in metals (Twerenbold, 1988). Nevertheless, research into superconducting tunnel junctions (and superconducting granules: Valette and Waysand, 1977; Seidel *et al.*, 1987) is likely to continue apace in the next few years.

In both superconductor and calorimeter detectors, the thickness of the X-ray absorbing layer is constrained by the need for good quantum efficiency at the X-ray energies of interest. The GSFC/Wisconsin group have proposed a silicon-based calorimeter for the US AXAF observatory (section 1.5) whose area is $0.5 \times 0.5 \, \text{mm}^2$ and whose thickness is 25 μm (fig. 6.2a and b). The latter figure represents the 1/e absorption depth for 6 keV X-rays in silicon. Similarly, Coron *et al.* (1985), having demonstrated the thermal detection of α-particles, have proposed a composite X-ray bolometer with a diamond absorber volume $1 \times 1 \times 0.25 \, \text{mm}^3$ (fig. 6.2c).

In the case of the AXAF detector, described in detail in section 6.3, the total heat capacity at 0.1 K should be (Moseley *et al.*, 1984):

$$C = 5.8 \times 10^{-15} \, \text{J/K}$$

which, with a ξ value of 2.56 (eq. 6.1) would lead to a fwhm energy resolution:

$$\Delta E = 1.1 \, \text{eV}$$

The calorimeter is thus, in principle, the ideal X-ray spectrometer, combining the high quantum efficiency of a proportional counter, Si(Li) detector or CCD with the excellent energy resolution of a Bragg crystal (section 1.3) or diffraction grating (section 1.5). At the time of writing (late 1987) it is not yet possible to say how closely the ultimate resolution of real calorimeters will approach the thermodynamic limit. One can only describe the experimental progress made so far and summarise recent theoretical studies of the non thermodynamic sources of energy blur dismissed in the derivation of eq. (6.1).

The first calorimeter constructed by the GSFC/Wisconsin group (McCammon *et al.*, 1984) achieved an energy resolution at 0.3 K

$$\Delta E = 270 \, \text{eV}$$

when illuminated by X-rays from an ^{55}Fe source mounted within the same ^{3}He cryostat. The heat capacity of this device ($0.25 \times 0.25 \, \text{mm}^2$ Si doped with antimony, compensated with boron and mounted on brass leads) was, however, rather high – around $10^{-11} \, \text{J/K}$. The measured resolution agreed rather well with expectation, once the effects of preamplifier noise and non-optimal signal filtering had been taken into account.

A second series of tests (Moseley *et al.*, 1985) used $0.2 \times 0.2 \times 0.8 \, \text{mm}^3$ samples of germanium, supported in a variety of ways – by brass, quartz or superconductor-coated carbon fibre leads. These devices began to match the energy resolution of conventional Si(Li) or CCD detectors (fig. 4.3) at a level

$$\Delta E = 130 \, \text{eV}$$

at an operating temperature of 0.3 K. The width of the baseline noise pulse height distribution, measured by sampling the detector output between X-ray events, was essentially the same as that of the signal peaks, indicating satisfactory thermal operation of the detector.

The third stage in the GSFC/Wisconsin programme returned to silicon as the absorbing material, but with an ion-implanted thermistor now occupying a small fraction of the total chip volume, in order to minimise the total heat capacity. Operated at 0.1 K, the baseline noise of these detectors was only 11 eV fwhm, in good agreement with the expected thermodynamic noise level of 8 eV. The width of the signal peak, however, was much broader (~ 130 eV), indicating for the first time the presence of a signal-dependent 'conversion noise' source.

Possible non-thermal sources of energy blur had been considered by Moseley et al. (1984) right at the outset of microcalorimeter development. Experiments with silver and gold X-ray conversion layers showed that, of these, charge trapping onto defect sites within the semiconductor crystal was responsible for the poor resolution of the ion-implanted silicon detectors. Use of a semi-metal (bismuth), with no trapping sites and low heat capacity, as the X-ray conversion medium then produced an energy resolution

$$\Delta E = 38 \, \text{eV}$$

with the device mounted in a ^{3}He/^{4}He 'dilution refrigerator' operating at 0.1 K (Holt et al., 1985).

Figs. 6.3b–d represent the current status of the GSFC/Wisconsin development programme (McCammon et al., 1987). An energy resolution of

$$\Delta E = 17 \, \text{eV}$$

has been achieved with ^{55}Fe illumination of the device illustrated in fig. 6.3a, at a bath temperature of 0.08 K. The thermalisation medium is now the semiconductor mercury cadmium telluride, with a Debye temperature of only 140 K. Comparing the output pulse traces of fig. 6.3b and fig. 6.1 indicates the progress which has been made in optimising calorimeter performance.

The work of other research groups in the field is generally at an earlier stage of development than that of the GSFC/Wisconsin collaboration. In the USA, groups at the Harvard-Smithsonian Center for Astrophysics and Bell Laboratories are aiming to produce an imaging bolometer (section

6.2.2), but have not so far made any low-temperature measurements. In the UK, a collaboration between MSSL and Queen Mary College, London, calling on existing IR bolometer expertise in the latter institution, is similarly just getting under way. In Europe, a consortium led by the Laboratoire de Physique Stellaire et Planetaire (LPSP) in France is developing a composite bolometer (composite in that, just as in fig. 6.3a, the energy absorber and temperature transducer are made from different materials, separately optimised: see fig. 6.2c), originally used in IR astronomy, as a particle spectrometer (Coron *et al.*, 1985a) and astronomical X-ray detector (Programme Asterix: Coron *et al.*, 1985b). In the LPSP design, diamond is the absorbing material and gallium-doped germanium, the thermistor. Diamond has an extremely high Debye temperature, and hence extremely low specific heat (Fiorini and Niinikoski, 1984). The speed of sound is also relatively high in diamond, which reduces the thermal time constant of the absorbing mass, improves the count rate capability of the detector and minimises the effects of thermal non-uniformity (below). The best X-ray energy resolution so far obtained at LPSP is

$$\Delta E = 2800\,\text{eV}$$

for 22 keV illumination of a relatively large (4 mm diameter) detector operated at the relatively high temperature of 0.5 K (Gabriel, 1987).

Alongside the experimental work, recent theoretical studies have tried to assess the limits which non-thermal effects may place on calorimeter resolution. Andersen (1986), while carefully considering energy loss mechanisms in diamond particle spectrometers, has noted that diamond is an efficient scintillator, with peak emission in the green part of the visible spectrum. This factor may also limit the ultimate energy resolution of composite X-ray bolometers. Fraser (1987) has derived limiting resolutions due to thermal non-uniformity in each of the three calorimeter designs shown in fig. 6.2. As first pointed out by Moseley *et al.* (1984), the shape of the voltage waveform produced at any thermistor output must depend on the position of energy deposition within the extended absorber. If an X-ray is absorbed close to a thermal link, its energy will flow rather more rapidly to the heat bath than if it is absorbed in the centre of the detector. Moseley *et al.* (1984) suggested the multiple thermistor design of fig. 6.2b in order to minimise any variation in estimated energy with position.

The contribution of the thermal equilibration effect to the detector's energy resolution is determined by the ratio, g, of the absorber time constant

$$\tau_{abs} = C/G_{abs}$$

to the time constant of the thermal links

$$\tau_{link} = C/G_{link}$$

G_{abs} and G_{link} are, respectively, the thermal conductances (in watts per kelvin) of the absorber and of the thermal links connecting the absorber mass to its heat bath. Appropriate solutions of the heat conduction equation allow $\Delta E/E$ to be estimated as a function of g (Fraser, 1987). Fig. 6.4 shows that g must in fact be made of order 10^{-3} if calorimeter resolution is (i) to be made competitive with that of an optimised Bragg

Fig. 6.3. (a) GSFC/Wisconsin silicon/aluminium/HgCdTe calorimeter (McCammon *et al.*, 1987). The silicon slab contains an ion-implanted thermistor. X-rays are absorbed in the HgCdTe chip. (b) 0.08 K pulse height spectrum from calorimeter of (a). The peaks at the right hand side are 5.9 and 6.4 keV Mn Kα and Kβ X-rays from an ^{55}Fe source. The peak at the left is the system noise, obtained by randomly sampling the steady state voltage level. The inset shows a typical output voltage pulse. (c) System noise on expanded scale. (d) Mn Kα peak on expanded scale. The discrepancy between baseline and signal widths may be due to non-uniform thermal response of the detector and/ or residual thermalisation noise. (*Courtesy D. McCammon.*)

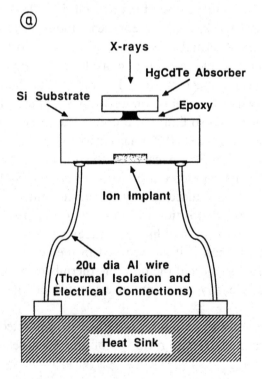

Fig. 6.3. (Continued)

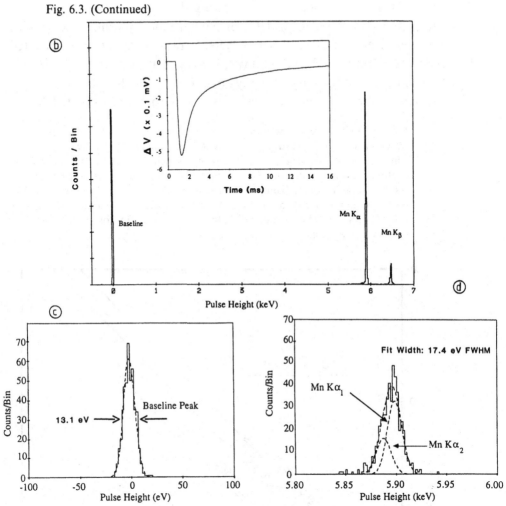

spectrometer, and (ii) approach the thermodynamic limit expressed in eq. (6.1). Since τ_{link} cannot be increased without limit without compromising the detector's count rate capability (a maximum value of $\sim 300\ \mu s$ is likely) and absorber time constants are likely to lie in the 1 μs range, thermal non-uniformity could potentially be a serious source of energy blur, limiting the designs of figs. 6.2a and b to around 20 eV and 6 eV, respectively, at an X-ray energy of 6 keV.

6.2.2   Imaging

The focal plane coverage of any conceivable microcalorimeter for astronomy is, as described above, severely limited by the need to minimise heat capacity. As in the case of CCDs (section 4.6.3) and other solid state detectors, mosaicing, the close packing of an array of individual detectors,

has been suggested as a means of allowing thermal 'imaging' of extended
sources and also of reducing detector background by using pixel to pixel
coincidence. No results with calorimeter arrays have yet been reported.
Programme Asterix at LPSP (Gabriel, 1987) has as its goal the production
of 3×3, then 8×8, bolometer arrays operating at 0.1 K.

A second form of imaging device is about to be investigated by the

Fig. 6.4. Fwhm energy resolution as a function of time constant ratio g.
Models a–c correspond to (a)–(c) of fig. 6.2. Illumination is by 5.9 keV X-rays.
The horizontal lines represent (i) the energy resolution attainable using a conventional Si(Li) detector or CCD (Chapter 4) at this energy, (ii) the energy
resolution (4 eV) attainable using a particular LiF crystal spectrometer around
6 keV (Byrnak et al., 1985), and (iii) 0.1 K thermodynamic energy resolution in
calorimeters a, b. The corresponding thermodynamic limit for model c is 4 eV
(Coron et al., 1985a).

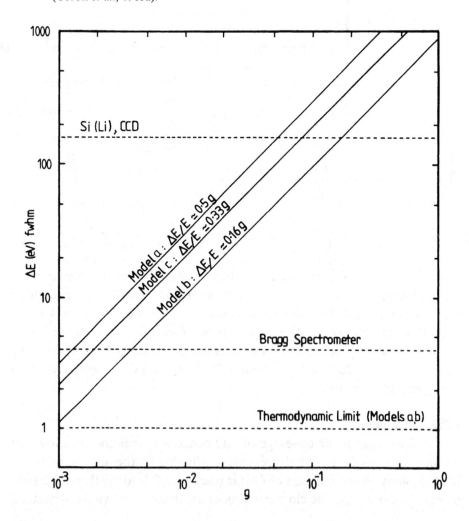

Harvard-Smithsonian/Bell Laboratories collaboration. These researchers intend to utilise the thermal non-uniformity of a single absorber – the variation of pulse rise-time and pulse height with X-ray absorption position – to produce the thermal analogue of a uniform resistive anode position encoder. Instead of four charge-collecting electrodes (figs. 2.20b and c), a position-sensitive calorimeter would have four matched thermistors either central to the sides of a square absorber, or at its vertices (fig. 6.2b).

Analysis of such a device brings out the fact that good spatial resolution Δx would require a large degree of thermal non-uniformity and would therefore have to be traded against energy resolution. In fact (Fraser, 1987):

$$\Delta x/L = R/g$$

where L is the linear size of the calorimeter and R is its noise-to-signal ratio. Provided system noise could indeed be reduced to the 1 eV level, detectors combining thermodynamic energy resolution and a few 6 keV image pixels per axis might be constructed. Such detectors would be obvious focal plane candidates for future missions with the high angular resolution of AXAF.

6.3   In-orbit realisation: the AXAF X-ray Spectroscopy Experiment

Fig. 6.5 (Holt *et al.*, 1985) is a schematic drawing of the GSFC/ Wisconsin X-ray Spectroscopy Experiment (XRS) for the AXAF observatory. The experiment contains, in addition to a segmented Si(Li) detector of the BBXRT design (section 4.3), the only X-ray calorimeter so far selected for any satellite mission. The physical characteristics of the absorber mass itself (figs. 6.2a, b) have already been discussed in some detail; this section considers the technical difficulties associated with operating such a device in orbit.

The most serious problem is one of cooling. First of all, the ^{3}He/^{4}He dilution refrigerators used to produce 0.1 K temperatures in the laboratory will not work in the microgravity of space. An adiabatic demagnetisation refrigerator (ADR) has therefore been adopted as the final cooling stage in a multistage cryostat for the XRS calorimeter. The ADR uses as its heat sink a ^{4}He cryostat at a temperature of 1.5 K, whose outer shell is in turn maintained at 65 K by a Stirling cycle mechanical cooler, a closed cycle heat engine incorporating two interconnected pistons (section 4.3). The Stirling cooler also directly cools the Si(Li) spectrometer which, in a payload context, acts as a long-lived, lower resolution backup to the XRS calorimeter. The lifetime of the calorimeter will be limited to around two years by boil-off of its liquid helium.

Adiabatic demagnetisation refrigerators work by the cyclic reorientation of the dipoles of a paramagnetic salt. Initially, the dipoles are randomly aligned and the salt is thermally coupled to the helium heat bath. A strong magnetic field (several tesla) is then applied. The dipoles align themselves to the field. Finally, with the salt 'pill' decoupled from the bath and with the field switched off, the dipoles randomly reorientate themselves, drawing energy from the salt, which cools to the desired 0.1 K.

Simple in concept, ADR based cryostats, which have also been considered for ESA's XMM mission (Briel *et al.*, 1988) may prove difficult and expensive to develop.

A second major area of concern is the provision of an X-ray window for the calorimeter mass. No laboratory measurements of any kind have yet been reported with the radiation source external to the calorimeter cryostat. In space, a window arrangement is needed which efficiently transmits X-rays but absorbs or reflects all other electromagnetic radiation, even the infrared emission from the spacecraft structure and the sky. The concept proposed for the XRS (fig. 6.5), and also by the LPSP consortium (Coron *et al.*, 1985b), is to have several metallised plastic shields anchored to the various cryostat shells. The outermost window (at a temperature of

Fig. 6.5. Schematic view of the AXAF X-ray spectroscopy experiment (Holt *et al.*, 1985). The experiment occupies one quadrant of the AXAF focal plane module. (*Courtesy S. S. Holt.*)

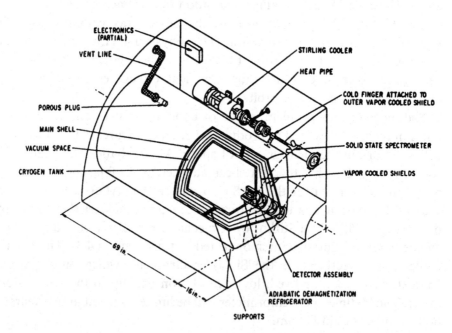

65 K) would have to be larger than the innermost (at 0.1 K) to allow for the conical nature of the X-ray beam converging from the X-ray optics. The transmission of the multiple window would, of course, fix the lower energy cutoff of the calorimeter bandpass to ~ 0.5 keV. Since calorimeters are 'constant ΔE' spectrometers, however, there must in any case be some lower energy limit below which they become uncompetitive, in terms of energy resolution, with 'constant $\Delta\lambda$' grating spectrometers. Comparing the 1 eV thermodynamic resolution of the proposed AXAF calorimeter with the expected performance of the AXAF low-energy transmission grating spectrometer ($\Delta\lambda = 0.06$ Å: Brinkman *et al.*, 1985), the cross-over point in energy resolution in fact occurs at an X-ray energy of about 0.5 keV.

It is not, at the present time, possible to conclude an account of X-ray calorimeters in any definitive way. Questions seem more appropriate. Can the thermodynamic limit to energy resolution be reached in real materials? Can Stirling cycle coolers be made to operate reliably for several years in space? (This latter question is, of course, also relevant to the deployment in space of all solid state X-ray detectors.) Can electromagnets be constructed which produce the required fields in the paramagnetic salt of an ADR without imparting excessive heat to its helium reservoir? Can appropriate cryogenic X-ray windows be constructed?

If the answer to any one of the last three questions is 'no', then the X-ray calorimeter may, like the NEADs described in section 5.4, be a detector whose laboratory promise is never translated into a working satellite instrument. If, on the other hand, the answer to all these questions is 'yes', X-ray astronomers of the near future may have at their disposal a spectrometer of quite revolutionary power.

6.4 Applications

Irrespective of whether or not the technical problems outlined in the previous section are ever overcome and calorimeters fulfil their potential in X-ray astronomy, they will certainly find application in a significant number of ground-based experiments (Table 6.1), all of which are right at the forefront of modern physics.

The proposed calorimeters divide between small-volume, high-resolution devices, similar to those needed in astronomy, and much larger, lower-resolution devices for particle searches. The final entry of Table 6.1 (with superior a) utilises a unique property of the calorimeter; its sensitivity to non-ionising events. Searches for the postulated heavy component of the non-luminous 'missing mass' in the universe, or for other 'exotic' particles, would measure the temperature rise directly induced in a suitable detector mass by recoil of the target nucleus, with no intervening ionisation stage in the detection process.

Table 6.1. *Proposed applications of single photon, single particle calorimeters*

Application	Comments
Ion beam analysis (S) Overley *et al.*, 1985; Andersen, 1986)	Calorimeter energy resolution ~20 times better than conventional surface barrier detectors
Electron neutrino mass measurement (S) (Coron *et al.*, 1985a; Stroke *et al.*, 1986)	v mass inferred from energy spectrum of electron capture β decay from implanted ^{163}Ho
Direct neutrino detection (L)	Large (100–10 000 kg) subdivided mass of Si. Temperature rise due to recoil of electron following elastic scattering of v
(Cabrera *et al.*, 1985)	
Double-beta decay (S) (Fiorini and Niinikoski, 1984)	Search for rare decay $(A,Z) \rightarrow (A,Z+2) + 2e^-$
Detection of heavy particle (photino) component of the galactic dark matter (L)^a (Smith, 1986)	Large (10 kg) subdivided mass of Si or other materials. Direct photon excitation by nuclear recoil

^{a} Utilises sensitivity of calorimeter to non-ionising events.
S = small-volume calorimeter; L = large-volume calorimeter.

Observational X-ray astronomy: a bibliography

Historical development

The following are among the many reviews which encapsulate the observational status of cosmic X-ray astronomy in the period 1962–78:

Friedman, H., Byram, F. T. and Chubb, T. A. *Science* 156 (1967) 374

Morrison, P. *Ann. Rev. Astron. Astrophys.* 5 (1967) 325

Giacconi, R., Gursky, H. and Van Speybroeck, L. P. *Ann. Rev. Astron. Astrophys.* 6 (1968) 373

Giacconi, R., Reidy, W. P., Vaiana, G. S., Van Speybroeck, L. P. and Zehnpfennig, T. F. *Space Sci. Rev.* 9 (1969) 3

Adams, D. J. *Contemp. Phys.* 12 (1971) 471

Giacconi, R. and Gursky, H., eds. *X-ray astronomy*, Astrophys. Space Sci. Library no. 43, D. Reidel, Dordrecht (1974)

Willmore, A. P. *Q. J. R. Astr. Soc.* 17 (1976) 400

Culhane, J. L. *Vistas in Astronomy* 19 (1977) 1

The modern era

Post-Einstein, post-EXOSAT it is no longer possible to review observational X-ray astronomy as a single entity. The first Einstein Observatory results appeared in *Astrophysical Journal Letters*, vol. 234, in 1979: the papers in that special journal issue, and in the collection edited by R. Giacconi and G. Setti (*X-ray astronomy*, Proc. NATO advanced study institute, Erice, Sicily, 1979; D. Reidel, Dordrecht (1980)), illustrate the separate classes of X-ray emitting objects which are now commonly the subjects of dedicated reviews. X-ray astronomy in the EXOSAT era is similarly represented in *Space Science Reviews*, vol. 40, (1985) (A. Peacock, ed.).

X-ray emission from normal stars: Rosner, R., Golub, L. and, Vaiana, G. S. *Ann. Rev. Astron. Astrophys.* 23 (1985) 413

Hot white dwarfs: Heise, J. *Space Sci. Rev.* 40 (1985) 79

Cataclysmic variables: Mason, K. *Space Sci. Rev.* 40 (1985) 99

Supernova remnants: Holt, S. S. In *Supernova Remnants and their X-ray Emission*, IAU symp. no. 101, P. Gorenstein and J. Danziger, eds., D. Reidel, Dordrecht, (1983); Raymond, J. C. *Ann. Rev. Astron. Astrophys.* 22 (1984), p. 75; Aschenbach, B. *Space Sci. Rev.* 40 (1985) 447

X-ray binaries: Joss, P. C. and Rappaport, S. *Ann. Rev. Astron. Astrophys.* 22 (1984) 537; Hayakawa, S. *Phys. Reports* 121 (1985) 1; White, N. E. and Mason, K. *Space Sci. Rev.* 40 (1985) 167.

Optical identification of galactic X-ray sources: Bradt, H. V. and McLintock, J. E. *Ann. Rev. Astron. Astrophys.* 21 (1983) 13

Active galactic nuclei: Giacconi, R. *Space Sci. Rev.* 30 (1981) 3; Balik, B. and Heckman, T. *Ann. Rev. Astron. Astrophys.* 20 (1982) 431; McHardy, I. *Space Sci. Rev.* 40 (1985) 559

Clusters of galaxies: Forman, W. and Jones, C. *Ann. Rev. Astron. Astrophys.* 20 (1982) 547; Fabian, A. *Space Sci. Rev.* 40 (1985) 653; Sarazin, C. L. *Rev. Mod. Phys.* 58 (1986) 1

The X-ray background: Giacconi, R. *Space Sci. Rev.* 30 (1981) 3; Boldt, E. *Phys. Reports* 146 (1987) 1

X-ray spectroscopy: Holt, S. S. *Vistas in Astronomy* 24 (1982) 301

Emission mechanisms: Pringle, J. E. *Ann. Rev. Astron. Astrophys.* 19 (1981) 137; Frank, J., King, A. R. and Raine, D. J. *Accretion Power in Astrophysics*, Cambridge University Press (1985)

X-ray data analysis techniques

The essential element of most astronomical X-ray data analysis is model-fitting. Whether one is dealing with an energy spectrum, a time-series or a sky map, one invariably selects a physically appropriate model of the emitting source and compares its predicted count rates with measured count rates, taking due account of the imperfect instrumental efficiency and resolution (energy, temporal or spatial). Model parameters (source temperature or spectral index, orbital period, position ...) are adjusted until the 'best fit' of the model to the data is obtained. A rigorous account of the theory of such 'composite hypothesis testing' is given by Lampton *et al.* (1976b). In X-ray astronomy, the most commonly used measure of the significance of the final fit is Pearson's χ^2 statistic. If D_i is the number of counts observed in the ith (space, time or energy) bin and F_i are the accumulations predicted by a model with p adjustable parameters, then the minimum value of the quantity

$$S = \sum_i^N (D_i - F_i)/\sigma_i^2 : \sigma_i^2 = F_i$$

found by a 'grid search' in parameter space or otherwise, is distributed as the χ^2 statistic with $N - p$ degrees of freedom (Martin, 1971).

Other techniques for the production of a 'point estimate' (the array of best-fit model parameters) and its associated confidence intervals (the regions of parameter space within which, to some degree of confidence, the best-fit values could lie) include the maximum likelihood method (Cash, 1979).

Pulse height analysis

Since conventional proportional counters (section 2.2.4) have rather modest energy resolution, three simple source models have been traditionally used in astronomical spectral fitting. In each case (Zombeck, 1982) we may write the predicted flux arriving at the detector in the form:

$$F(E) = C\exp(-\sigma_H(E)N_H)f(s,E) \text{ photons/cm}^2 \text{ s keV}$$

where C is a normalisation constant (the first adjustable parameter), N_H is the hydrogen column density (atoms/cm^2) along the line of sight to the source (adjustable parameter 2), and σ_H is the photoelectric cross-section per hydrogen atom. The exponential term thus represents absorption in the interstellar medium and in gas clouds local to the emitting region. $f(s, E)$ describes the intrinsic spectral shape of the source, controlled by the third adjustable parameter s.

For thermal Bremmstrahlung emission from an optically thin hot gas, s corresponds to the temperature of the source, T, and

$$f = \bar{g}\exp(-E/kT)/E(kT)^{\frac{1}{2}}$$

$\bar{g}$ is the temperature-averaged Gaunt factor (Greene, 1959) which corrects the spectrum for quantum mechanical effects. k is the Boltzmann constant.

For power law emission, often associated with a synchrotron origin of the X-rays, s coincides with the spectral index, n, and

$$f = E^{-n}$$

For a black body emitter, s is again the source temperature and

$$f = E^2/\{\exp(-E/kT)-1\}$$

The presence of line emission at a single X-ray energy E_0 may be simulated by the addition of a Gaussian term:

$$B\exp(-(E-E_0)^2/2\sigma_0^2): B, \sigma_0 \text{ constants}$$

to any of these expressions. A fourth type of source model, becoming more useful with the development of detectors with better energy resolution, estimates f numerically from a computer model of the ionisation within an optically thin plasma (Raymond and Smith, 1977; Mewe and Gronenschild, 1981). Here, the plasma temperature and the elemental abundances (O, Si, S, Ar, Ca, Fe, etc.) are the adjustable parameters.

Conventionally then, one assumes a form for the source spectrum $F(E)$, folds it through the instrumental response, and compares predicted and measured detector count rates. Alternatively, one may derive source photon spectra without prejudice as to the underlying mechanisms by deconvolving the instrumental response from the measured count rates, using, for example, the maximum entropy method (Kayat *et al.*, 1980; Willingale, 1981; Narayan and Nityananda, 1986) or direct spectral restoration techniques (Blissett and Cruise, 1979).

Temporal analysis

Cosmic X-ray sources are variable on many timescales. The simplest form of temporal analysis, therefore, is χ^2 testing against the hypothesis of steady flux to establish rigorously whether or not apparent variations in measured time-series are statistically significant (Marshall *et al.*, 1981).

At a higher level of sophistication, the commonest form of temporal analysis is period-searching: the most powerful technique for abstracting periodicities from time-series data is Fourier analysis. The application of the fast Fourier transform (FFT) to X-ray data, however, is far from straightforward. The ideal input to the FFT is a series of intensity samples obtained at regularly spaced intervals. Satellite observations, particularly from low earth orbit, inevitably contain irregular gaps because of earth obscuration, telemetry losses and spacecraft passage through regions of high background such as the South Atlantic Anomaly. Ponman (1981) describes how this problem of incomplete coverage can be overcome.

If a modulation of frequency ν is present in the data, the source power spectrum $P(\nu)$ (the power spectral density is defined as the sum of the squares of the Fourier sine and cosine coefficients) will contain a sharp peak centred on ν. Aliases – spurious peaks which are artefacts of the sampling and analysis techniques – and the higher harmonics of ν may also be present (Ponman, 1981).

A simpler form of period-searching easily applied to incomplete data sets uses epoch folding. Here, the phase of each data point is established relative to some arbitrary zero and hypothesised period, the period being adjusted until the best fit between the data and the model sinusoid is obtained.

Very rapid chaotic variability or 'flickering', such as that observed in the black hole candidate Cygnus X-1, while correlated on timescales of a few tenths of a second, definitely does not represent a coherent phenomenon and must be analysed in terms of specialised shot-noise models (Sutherland *et al.*, 1978) which consider the emission as a series of random bursts ('shots') of well-defined shape.

Spatial analysis

The production of sky maps from scanning proportional counter data is discussed in section 1.3, with reference to the Ariel V Sky Survey Instrument. With imaging instruments, there are two distinct types of problem:

(i) estimating the position of weak sources (represented by a handful of counts) in the presence of background;

(ii) producing maps of extended sources, or of variations in the diffuse X-ray background.

The former problem is usually addressed by the use of automated search routines which step a test cell through the array of image pixels and compare the number of counts in that cell with the local background level, estimated from the pixels bordering the cell. Approximate locations thus identified (or identified visually) can then be refined by maximum likelihood fitting of the known instrumental point response (representing the combined blur of the X-ray optics and detector) to the observed distribution of photon positions (Cash, 1979).

Methods for image restoration which produce 'smoothed' maps which are as free as possible from the effects of instrumental and statistical degradation are discussed by Daniell (1984) and, in the special context of coded aperture imaging, by Willingale *et al.* (1984). Iterative methods include:

(i) The maximum entropy method (MEM), mentioned above in the context of pulse height analysis, which seeks to maximise the configurational entropy

$$-\sum_i f_i \ln(f_i)$$

of the solution (f_i is the estimated flux in the ith image pixel) while at the same time minimising a fit statistic such as χ^2 between the estimated and measured pixel count rates (Willingale, 1981; Ponman, 1984).

(ii) The algebraic reconstruction technique (ART) (Stevens and Garmire, 1973).

(iii) Bayesian conditional probability methods (Willingale *et al.*, 1984).

The most commonly used non-iterative filter is the minimum mean square error or Wiener filter (Willingale *et al.*, 1984).

References

Ables, J. G. *Proc. Astr. Soc. Aust.* 1 (1968) 172

Adams, D. J., Janes, A. F. and Whitford, C. H. *Astron. Astrophys.* 20 (1972) 121

Adams, G. P., Rochester, G. K., Sumner, T. J. and Williams, O. R. *J. Phys. E* 20 (1987) 1261

Adams, J. *Adv. Electron. Electron Phys.* 22A (1966) 139

Adams, J. and Manley, B. W. *IEEE Trans. Nucl. Sci.* NS-13 (1966) 88

Ainbund, M. R. and Maslenkov, I. P. *Instr. Exp. Tech.* 26 (1983) 650

Aitken, D. W. *IEEE Trans. Nucl. Sci.* NS-15 (1968a) 10

Aitken, D. W. *IEEE Trans. Nucl. Sci.* NS-15 (1968b) 214

Aitken, D. W., Beron, B. L., Yenicay, G. and Zullinger, H. R. *IEEE Trans. Nucl. Sci.* NS-14 (1967) 468

Albridge, R. G. *et al. Nucl. Instr. Meth.* B 18 (1987) 582

Alig, R. C., Bloom, S. and Struck, C. W. *Phys. Rev. B* 22 (1980) 5565

Alkhazov, G. D., Komar, A. P. and Vorobjev, A. A. *Nucl. Instr. Meth.* 48 (1967) 1

Alkhazov, G. D. *Nucl. Instr. Meth.* 89 (1970) 155

Allemand, R. and Thomas, G. *Nucl. Instr. Meth.* 137 (1976) 141

Allington-Smith, J. R. and Schwarz, H. E. *Q. J. R. Astr. Soc.* 25 (1984) 267

Andersen, H. H. *Nucl. Instr. Meth.* B 15 (1986) 722

Anderson, D. F. *IEEE Trans. Nucl. Sci.* NS-28 (1981) 842

Anderson, D. F. *IEEE Trans. Nucl. Sci.* NS-32 (1985) 516

Anderson, D. F., Ku, W., Mitchell, D. D., Novick, R. and Wolff, R. S. *IEEE Trans. Nucl. Sci.* NS-24 (1977a) 283

Anderson, D. F., Bodine, O. H., Novick, R. and Wolff, R. S. *Nucl. Instr. Meth.* 144 (1977b) 485

Anderson, K. A. *Space. Sci. Rev.* 13 (1972) 337

Andresen, R. D., Leimann, E. A., Peacock, A. and Taylor, B. G. *Nucl. Instr. Meth.* 146 (1977a) 391

Andresen, R. D., Leimann, E. A., Peacock, A., Taylor, B. G., Brownlie, G. and Sanford, P. *IEEE Trans. Nucl. Sci.* NS-24 (1977b) 810

Andresen, R. D. *et al. Space Sci. Rev.* 30 (1981) 243

Anger, C. D. *et al. Geophys. Res. Lett.* 14 (1987) 387

Anger, H. O. *Rev. Sci. Instrum.* 29 (1958) 27

Armantrout, G. A., Swierkowski, S. P. and Yee, J. H. *IEEE Trans. Nucl. Sci.* NS-24 (1977) 121

Armitage, J., Waterhouse, J. and Wong, L. *Nucl. Instr. Meth. A* 273 (1988) 476

Arndt, U. *Nucl. Instr. Meth.* 201 (1982) 13

Asam, A. R. *Opt. Eng.* 17 (1978) 640

Aschenbach, B. *Rep. Prog. Phys.* 48 (1985) 579

Aschenbach, B. *et al.* 'Lyman assessment study', ESA SCI(85)4 (1985)

Aschenbach, B., Briel, U. G., Pfeffermann, E., Brauninger, H., Hippmann, H. and Trümper, J. *Nature* 330 (1987a) 232

Aschenbach, B. *et al.* 'The high throughput X-ray spectroscopy mission: report of the telescope working group', ESA SP-1084 (1987b)

Authinarayanan, A. and Dudding, R. W. *Adv. Electron. Electron Phys.* 40A (1976) 153

Ayre, C. A., Bhat, P. N., Owens, A., Summers, W. M. and Thomson, M. G. *Phil. Trans. R. Soc.* 301 (1981) 687

Bailey, R. *et al. Nucl. Instr. Meth.* 213 (1983) 201

Bailey, T. A., Smith, A. and Turner M. J. L. *Nucl. Instr. Meth.* 155 (1978) 177

Balsiger, H. *et al. J. Phys. E* 20 (1987) 759

Bambynek, W. *et al. Rev. Mod. Phys.* 44 (1972) 716

Bardas, D., Kellogg, E., Murray, S. S. and Enck, R. *Rev. Sci. Instrum.* 49 (1978) 1273

Barstow, M. A., Fraser, G. W. and Milward, S. R. *Proc. SPIE* 597 (1985) 352

Barstow, M. A., Kent, B. J., Whiteley, M. J. and Spurrett, P. H. *J. Mod. Opt.* 34 (1987) 1491

Barstow, M. A. and Pounds, K. A. 'The Wide Field Camera for ROSAT: observing stars', *Proc. Nato Advanced Study Institute*, Cargese, Corsica (1987) (in press)

Bartl, W. and Neuhofer, G., eds. Proc. Third Wire Chamber Conf., Vienna, *Nucl. Instr. Meth.* 217 (1983)

Bartl, W., Neuhofer, G. and Regler, M., eds. Proc. Fourth Wire Chamber Conf., Vienna, *Nucl. Instr. Meth. A* 252 (1986)

Bateman, J. E. *Nucl. Instr. Meth.* 144 (1977a) 537

Bateman, J. E. *Nucl. Instr. Meth.* 142 (1977b) 371

Bateman, J. E. *Nucl. Instr. Meth.* 221 (1984) 131

Bateman, J. E. *Nucl. Instr. Meth. A* 240 (1985) 177

Bateman, J. E. and Apsimon, R. J. *Adv. Electron. Electron Phys.* 52 (1979) 189

Bateman, J. E., Apsimon, R. J. and Barlow, F. E. *Nucl. Instr. Meth.* 137 (1976) 61

Bateman, J. E., Apsimon, R. J. and Arndt, U. W. Rutherford Appleton Laboratory report RL 81-008 (1981)

Bateman, J. E., Connolly, J. F. and Stephenson, R. Rutherford Appleton Laboratory report RAL-82-080 (1982)

Bateman, J. E., Connolly, J. F. and Stephenson, R. Rutherford Appleton Laboratory report RAL-84-089 (1984)

Bateman, R. S. UK patent application GB 2020481A (1979)

Bates, C. W. *Adv. Electron. Electron. Phys.* 28A (1969) 451

Bauer, R. W. and Weingart, R. C. *IEEE Trans. Nucl. Sci.* NS-15 (1968) 147

Becchetti, F. D., Raymond, R. S., Ristinen, R. A., Schnepple, W. F. and
Ortale, C. *Nucl. Instr. Meth.* 213 (1983) 127

Bedford, D. K. and St. J. Mann, M. F. *J. Phys. E* 17 (1984) 866

Bichsel, H. *Nucl. Instr. Meth. A* 235 (1985) 174

Birks, J. B. *The Theory and Practice of Scintillation Counting*, Pergamon,
Oxford (1964)

Bjorkholm, P. J., Van Speybroeck, L. P. and Hecht, M. *Proc. SPIE* 106 (1977)
189

Bleeker, J. A. M., ed. *Proc. ESA Workshop 'Cosmic X-ray spectroscopy
mission'*, ESA SP-239 (1985)

Bleeker, J. A. M. and Overtoom, J. M. *Nucl. Instr. Meth.* 167 (1979) 505

Bleeker, J. A. M., Burger, J. J., Deerenburg, A. J. M., Scheepmaker, A.,
Swanenberg, B. N. and Tanaka, Y. *Astrophys. J.* 147 (1967) L391

Bleeker, J. A. M., Huizenga, H., den Boggende, A. J. F. and Brinkman, A. C.
IEEE Trans. Nucl. Sci. NS-27 (1980) 176

Blissett, R. J. and Cruise, A. M. *Mon. Not. R. Astr. Soc.* 186 (1979) 45

Bocciolini, M., Conti, A., Di Caporiacco, G., Parrini, G. and Quareni Vignu-
delli, A. *Nucl. Instr. Meth A* 240 (1985) 36

den Boggende, A. J. F., Brinkman, A. C. and de Graff, W. *J. Phys. E* 2 (1969)
179

den Boggende, A. J. F. and Lafleur, H. T. J. A. *IEEE Trans. Nucl. Sci.* NS-22
(1975) 555

den Boggende, A. J. F. and Schrijver, C. J. *Nucl. Instr. Meth.* 220 (1984) 561

Borer, K. *et al. Nucl. Instr. Meth. A* 253 (1987) 548

Borkowski, C. J. and Kopp, M. K. *Rev. Sci. Instrum.* 39 (1968) 1515

Borkowski, C. J. and Kopp, M. K. *IEEE Trans. Nucl. Sci.* NS-17 (1970) 340

Borkowski, C. J. and Kopp, M. K. *IEEE Trans. Nucl. Sci.* NS-19 (1972) 161

Borkowski, C. J. and Kopp, M. K. *Rev. Sci. Instrum.* 46 (1975) 951

Boutot, J. P., Delmotte, J. C., Miehe, J. A. and Sipp, B. *Rev. Sci. Instrum.* 48
(1977) 1405

Boutot, J. P., Eschard, G., Polaert, P. and Duchenois, V. *Adv. Electron.
Electron Phys.* 40A (1976) 103

Bowles, J. A., Patrick, T. J., Sheather, P. H. and Eiband, A. M. *J. Phys. E* 7
(1974) 183

Bowyer, S. *Rev. Sci. Instrum.* 36 (1965) 1009

Bowyer, S., Byram, E. T., Chubb, T. A. and Friedman, H. *Nature* 201 (1964a)
1307

Bowyer, S., Byram, E. T., Chubb, T. A. and Friedman, H. *Science* 146 (1964b)
912

Bowyer, S., Malina, R., Lampton, M., Finley, D., Paresce, F. and Penegor, G.
Proc. SPIE 279 (1981a) 176

Bowyer, S., Kimble, R., Paresce, F., Lampton, M. and Penegor, G. *Appl. Opt.*

20 (1981b) 477

Boyd, R. L. F., *Proc. Roy. Soc.* A366 (1979) 1

Boyle, W. and Smith, G. *Bell. Sys. Tech. J.* 49 (1970) 587

Bradt, H. *Adv. Space. Res.* 2 (1982) 315

Bradt, H., Garmire, G., Spada, G., Sreekantan, B. V. and Gorenstein, P. *Space Sci. Rev.* 8 (1968) 471

Bradt, H., Doxsey, R. and Jernigan, J. G. *Adv. Space Exploration* 3 (1979) 3

Breskin, A. *et al. Nucl. Instr. Meth.* 143 (1977) 29

Breskin, A., Charpak, G., Majewski, S., Melchart, G., Petersen, G. and Sauli, F. *Nucl. Instr. Meth.* 161 (1979) 19

Breskin, A. and Chechik, R. *IEEE Trans. Nucl. Sci.* NS-32 (1984) 504

Briel, U. G. and Pfeffermann, E. *Nucl. Instr. Meth.* A242 (1986) 376

Briel, U. G. *et al.* 'The high throughput spectroscopy mission: report of the instrument working group', ESA SP-1092 (1988)

Brinckmann, P. *Phys. Lett.* 15 (1965) 305

Brinkman, A. C., Heise, J. and de Jager, C. *Phillips Tech. Rev.* 34 (1974) 43

Brinkman, A. C., Dijkstra, J. H., Geerlings, W. F. P. A. L., Van Rooijen, F. A., Timmerman, C. and de Korte, P. A. J. *Appl. Opt.* 19 (1980) 1601

Brinkman, A. C. *et al. Proc. SPIE* 597 (1985) 232

Bronshtein, I. M., Yevdokimov, A. V., Stozharov, V. M. and Tyutikov, A. M. *Rad. Eng. Electron. Phys.* 24 (1979) 150

Brooks, F. C. *Nucl. Instr. Meth.* 162 (1979) 477

Bross, A. D. and Clegg, D. B. *Nucl. Instr. Meth. A* 247 (1986) 309

Buchanan, R. A. and Wickersheim, K. A. *IEEE Trans. Nucl. Sci.* NS-15 (1968) 95

Burger, A., Shilo, I. and Schieber, M. *IEEE Trans. Nucl. Sci.* NS-30 (1983) 368

Burton, W. M. *Adv. Space. Res.* 2 (1982) 221

Burton, W. M. and Powell, A. B. *Appl. Opt.* 12 (1973) 87

Byrnak, B., Christensen, F. E., Westergaard, N. J. and Schnopper, H. W. *Proc. SPIE* 597 (1985) 309

Cabrera, B., Krauss, L. M. and Wilczek, F. *Phys. Rev. Lett.* 55 (1985) 25

Campbell, J. L. and Ledingham, K. W. D. *Brit. J. Appl. Phys.* 17 (1966) 769

Canali, C., Martini, M., Ottaviani, G. and Alberigi Quaranta, A. *IEEE Trans. Nucl. Sci.* NS-19 (1972) 9

Canizares, C., Clark, G., Bardas, D. and Markert, T. *Proc. SPIE* 106 (1977) 154

Canizares, C., Schattenburg, M. L. and Smith, H. I. *Proc. SPIE* 597 (1985a) 253

Canizares, C., Markert, T. and Clark, G. W. *Proc. SPIE* 597 (1985b) 241

Carnes, J. and Kosonocky, W. *RCA Rev.* 33 (1972) 327

Carpenter, G. F., Coe, M. J., Engel, A. R. and Quenby, J. J. *Proc. Roy. Soc. A* 350 (1976) 521

Carruthers, G. R. *Appl. Opt.* 26 (1987) 2925

Carter, J. N. *et al. Nucl. Instr. Meth.* 196 (1982) 477

Cash, W. *Astrophys. J.* 228 (1979) 939

Catchpole, C. E. and Johnson, C. B. *Publ. Astron. Soc. Pacific* 84 (1972) 134

Catura, R. C. and Smithson, R. C. *Rev. Sci. Instrum.* 50 (1979) 219

Catura, R. C., Joki, E. G., Bakke, J. C., Rapley, C. G. and Culhane, J. L. *Mon. Not. R. Astr. Soc.* 168 (1974) 217

Catura, R. C., Brown, W. A., Joki, E. G. and Nobles, R. A. *Opt. Eng.* 22 (1983) 140

Chalmeton, V. and Eschard, G. *Adv. Electron. Electron Phys.* 33A (1972) 167

Chappell, J. H. and Murray, S. S. *Nucl. Instr. Meth.* 221 (1984) 159

Charalambous, P. M. *et al. Nucl. Instr. Meth.* 221 (1984) 183

Charalambous, P. M., Dean, A. J., Lewis, R. A. and Dipper, N. A. *Nucl. Instr. Meth. A* 238 (1985) 533

Charles, M. W. *J. Phys. E* 5 (1972) 95

Charles, M. W. and Cooke, B. A. *Nucl. Instr. Meth.* 61 (1968) 31

Charpak, G. *Nature* 270 (1977) 479

Charpak, G. *Nucl. Instr. Meth.* 201 (1982) 181

Charpak, G. *Proc. SPIE* 597 (1985) 170

Charpak, G., Bouclier, R., Bressani, T., Favier, J. and Zupancic, C. *Nucl. Instr. Meth.* 62 (1968) 235

Chowanietz, E. G. PhD Thesis, University of Leicester (1986)

Chowanietz, E. G., Lumb, D. H. and Wells, A. A. *Proc. SPIE* 597 (1985) 381

Clampin, M. and Edwin, R. P. *Rev. Sci. Instrum.* 58 (1987) 167

Clark, G. W. *Phys. Rev. Lett.* 14 (1965) 91

Clark, G. W., Lewin, W. H. G. and Smith, W. B. *Astrophys. J.* 151 (1968) 21

Cline, D. B. *et al. IEEE Trans. Nucl. Sci.* NS-26 (1979) 3302

Cockshott, R. A. and Mason, I. M. *IEEE Trans. Nucl. Sci.* NS-31 (1984) 107

Cohen, D. *X-ray Spectrometry* 16 (1987) 237

Coleman, C. I. and Boksenberg, A. *Contemp. Phys.* 17 (1976) 209

Colson, W. B., McPherson, J. and King, F. T. *Rev. Sci. Instrum.* 44 (1973) 1694

Cooke, B. A., Pounds, K. A., Stewardson, E. A. and Adams, D. J. *Astrophys. J.* 150 (1967) L189

Cooke, B. A., Griffiths, R. E. and Janes, A. F. *Nucl. Instr. Meth.* 106 (1973) 147

Coron, N. *et al. Nature* 314 (1985a) 75

Coron, N. *et al. Proc. SPIE* (1985b) 389

Cortez, J. and Laprade, B. *Proc. SPIE* 427 (1982) 53

Crannell, C. J. *et al.* Goddard Space Flight Center preprint X682-77-121 (1977)

Cresswell, J., Carvahlo, I., LeNoble, M., Benolo, A. and Kule, R. *IEEE Trans. Nucl. Sci.* NS-33 (1986) 90

Crocker, J., Rafal, M., Paresce, F., Hiltner, A. and Denman, B. *Proc. SPIE* 627 (1986) 631

Csorba, I. P. *Appl. Opt.* 19 (1980) 3863

Culhane, J. L. and Fabian, A. C. *IEEE Trans. Nucl. Sci.* NS-19 (1972) 569

Culhane, J. L., Herring, J., Sanford, P. W., O'Shea, G. and Philips, J. *J. Sci. Instrum.* 43 (1966) 908

Curran, J. E. *J. Vac. Sci. Tech.* 14 (1977) 108

Curran, S. C. *Luminescence and the Scintillation Counter*, Butterworths, London (1953)

Curran, S. C. and Craggs, J. D. *Counting Tubes*, Butterworths, London (1949)

Cuzin, M. *Nucl. Instr. Meth. A* 253 (1987) 407

Dabrowski, A. J. *et al. IEEE Trans. Nucl. Sci.* NS-28 (1981) 536

Dabrowski, A. J. *Adv. X-ray Analysis* 25 (1982) 1

Dabrowski, A. J., Szymczyk, W. M., Iwanczyk, J. S., Kusmiss, J. H., Drummond, W. and Ames, L. *Nucl. Instr. Meth.* 213 (1983) 89

Damerell, C. *Nucl. Instr. Meth.* 226 (1984) 26

Daniell, G. J. *Nucl. Instr. Meth.* 221 (1984) 67

Davelaar, J., Manzo, G., Peacock, A., Taylor, B. G. and Bleeker, J. A. M. *IEEE Trans. Nucl. Sci.* NS-27 (1980) 196

Davelaar, J., Peacock, A. and Taylor, B. G. *IEEE Trans. Nucl. Sci.* NS-29 (1982) 142

Dean, A. J. *Nucl. Instr. Meth.* 221 (1984) 265

Dean, A. J. and Dipper, N. A. *Nucl. Instr. Meth.* 179 (1981) 147

Demarest, J. A. and Watson, R. L. *Nucl. Instr. Meth B* 24/25 (1987) 296

Desai, U. D. and Holt, S. S. *IEEE Trans. Nucl. Sci.* NS-19 (1972) 592

Dicke, R. H. *Astrophys. J.* 153 (1968) L101

Di Cocco, G., Dusi, W., Labanti, C., Spada, G. F., Ventura, G. and Siffert, P. *Proc. ESA workshop 'Cosmic X-ray spectroscopy mission'*, ESA SP-239 (1985) 233

Dolan, K. W. and Chang, J. *Proc. SPIE* 106 (1977) 178

Dotani, T. *et al. Nature* 330 (1987) 230

Duval, B. P., Bateman, J. E. and Peacock, N. J. *Rev. Sci. Instrum.* 57 (1986) 2156

Dyer, C. S., Trombka, J. I., Seltzer, S. M. and Evans, L. G. *Nucl. Instr. Meth.* 173 (1980) 585

Eberhardt, E. H. *IEEE Trans. Nucl. Sci.* NS-28 (1981) 712

Eberhardt, J. E., Ryan, R. D. and Tavendale, A. J. *Nucl. Instr. Meth.* 94 (1971) 463

Edgecumbe, J. and Garwin, E. L. *J. Appl. Phys.* 37 (1966) 3321

EEV CCD Imaging III, English Electric Valve Co. (1987)

Eichinger, P. *Nucl. Instr. Meth. A* 253 (1987) 313

Ellison, J., Hall, G. and Roe, S. *Nucl. Instr. Meth. A* 260 (1987) 353

Ellul, J. P., Tsoi, H., White, J. J., King, M. H., Bradley, W. C. and Colvin, D. W. *Proc. SPIE* 501 (1984) 117

Elvis, M., Fabricant, D. and Gorenstein, P. *Appl. Opt.* (1988) (in press)

Eng, W. and Landecker, P. B. *Nucl. Instr. Meth.* 190 (1981) 149

Epperson, P. M., Sweedler, J. V., Denton, M. B., Sims, G. R., McCurnin, T. W. and Aikens, R. S. *Opt. Eng.* 26 (1987) 715

Ewan, G. T. *Nucl. Instr. Meth.* 162 (1979) 75

Fabian, A. C., Willingale, R., Pye, J., Murray, S. S. and Fabbiano, G. *Mon. Not. R. Astr. Soc.* 193 (1980) 175

Fabjan, C. W. and Fischer, H. G. *Rep. Prog. Phys.* 43 (1980) 1003

Fano, U., *Phys. Rev.* 72 (1947) 26

Faruqi, A. *J. Phys. E* 8 (1975) 635

Faruqi, A., Huxley, H. E. and Kress, M. *Nucl. Instr. Meth. A* 252 (1986) 234

Fiorini, E. and Niinikoski, T. O. *Nucl. Instr. Meth.* 224 (1984) 1983

Firmani, C., Ruiz, E., Carlson, C. W., Lampton, M. and Paresce, F. *Rev. Sci. Instrum.* 53 (1982) 570

Fischer, J., Okuno, M. and Walenta, L. A. H. *Nucl. Instr. Meth.* 151 (1978) 451

Fischer, J., Radeka, V. and Smith, G. C. *Nucl. Instr. Meth. A* 252 (1986) 239

Fisher, P. C. and Meyerott, A. J. *IEEE Trans. Nucl. Sci.* NS-13 (1966) 580

Fishman, G. J. *Rev. Sci. Instrum.* 52 (1981) 1143

Fontaine, A., Dartyge, E., Jucha, A. and Tourillon, G. *Nucl. Instr. Meth. A* 253 (1987) 519

Ford, J. L. C. *Nucl. Instr. Meth.* 162 (1979) 277

Forman, W. *et al. Astrophys. J. (Suppl.)* 38 (1978) 357

Fraser, G. W. *Nucl. Instr. Meth.* 195 (1982) 523

Fraser, G. W. *Nucl. Instr. Meth.* 206 (1983a) 251

Fraser, G. W. *Nucl. Instr. Meth.* 206 (1983b) 445

Fraser, G. W. *Nucl. Instr. Meth.* 221 (1984) 115

Fraser, G. W. *Nucl. Instr. Meth. A* 228 (1985) 532

Fraser, G. W. *Nucl. Instr. Meth. A* 256 (1987) 553

Fraser, G. W. and Mathieson, E. *Nucl. Instr. Meth.* 179 (1981a) 591

Fraser, G. W. and Mathieson, E. *Nucl. Instr. Meth.* 184 (1981b) 537

Fraser, G. W. and Mathieson, E. *Nucl. Instr. Meth.* 180 (1981c) 597

Fraser, G. W. and Mathieson, E. *Nucl. Instr. Meth. A* 247 (1986) 544, 566

Fraser, G. W. and Pearson, J. F. *Nucl. Instr. Meth. A* 219 (1984) 199

Fraser, G. W., Mathieson, E., Evans, K. D., Lumb, D. H. and Steer, B. *Nucl. Instr. Meth.* 180 (1981a) 255

Fraser, G. W., Mathieson, E. and Evans, K. D. *Nucl. Instr. Meth.* 180 (1981b) 269

Fraser, G. W., Mathieson, E., Lewis, M. and Barstow, M. A. *Nucl. Instr. Meth.* 190 (1981c) 53

Fraser, G. W., Barstow, M. A., Whiteley, M. J. and Wells, A. *Nature* 300 (1982) 509

Fraser, G. W., Pearson, J. F., Smith, G. C., Lewis, M. and Barstow, M. A. *IEEE Trans. Nucl. Sci.* NS-30 (1983) 455

Fraser, G. W., Barstow, M. A., Pearson, J. F., Whiteley, M. J. and Lewis, M. *Nucl. Instr. Meth. A* 224 (1984) 287

Fraser, G. W., Pearson, J. F. and Lees, J. E. *Nucl. Instr. Meth. A* 256 (1987a) 401

Fraser, G. W., Pearson, J. F. and Lees, J. E. *Nucl. Instr. Meth. A* 254 (1987b) 447

Fraser, G. W., Barstow, M. A. and Pearson, J. F. *Nucl. Instr. Meth. A* 273 (1988a) 667

Fraser, G. W., Pearson, J. F. and Lees, J. E. *IEEE Trans. Nucl. Sci.* NS-35 (1988b) 529

Fraser, G. W., Whiteley, M. J. and Pearson, J. F. *Proc. SPIE* 597 (1985) 343

Friedman, H. *Ann. NY Acad. Sci.* 198 (1972) 267

Friedman, H. *Proc. Roy. Soc. A* 366 (1979) 423

Frontera, F. *et al. Nucl. Instr. Meth. A* 235 (1985) 573

Frost, K. J., Rothe, E. D. and Peterson, L. E. *J. Geophys. Res.* 71 (1966) 4079

Fulbright, H. W. *Nucl. Instr. Meth.* 169 (1979) 341

Gabriel, A., ed. Laboratoire de Physique Stellaire et Planetaire, Rapport d'acti-vite 1985–86, rapport G87-3 (1987)

Gao, R. S., Gibner, P. S., Newman, J. H., Smith, K. A. and Sieblings, R. F. *Rev. Sci. Instrum.* 55 (1984) 1756

Garcia, M. R., Grindlay, J. E., Burg, R., Murray, S. S. and Flanagan, J. *IEEE Trans. Nucl. Sci.* NS-33 (1986) 735

Garmire, G. P. *et al. Proc. SPIE* 597 (1985) 261

Gatti, E. and Rehak, P. *Nucl. Instr. Meth.* 225 (1984) 608

Gatti, E., Oba, K. and Rehak, P. *IEEE Trans. Nucl. Sci.* NS-30 (1983) 461

Gatti, E., Rehak, P. and Walton, J. T. *Nucl. Instr. Meth.* 226 (1984) 129

Gear, C. W. *Proc. Skytop Conf. USAEC Conf. no. 670301* (1969) 552

Gehrels, N. *et al. IEEE Trans. Nucl. Sci.* NS-31 (1984) 307

Germer, R. and Meyer-Ilse, W. *Rev. Sci. Instrum.* 57 (1986) 426

Gerritsen, H. C., van Brug. H. and van der Wiel, M. J. *J. Phys. E* 19 (1986) 1040

Giacconi, R. and Rossi, B. *J. Geophys. Res.* 65 (1960) 773

Giacconi, R., Gursky, H., Paolini, F. R. and Rossi, B., *Phys. Rev. Lett.* 9 (1962) 439

Giacconi, R., Harmon, N. F., Lacey, R. F. and Szilagyi, Z. *J. Opt. Soc. Am.* 55 (1965) 345

Giacconi, R., Gursky, H. and Van Speybroeck, L. P. *Ann. Rev. Astron. Astrophys.* 6 (1968) 373

Giacconi, R. *et al. Astrophys. J.* 230 (1979) 540

Gilmore, R. S. *et al. Nucl. Instr. Meth.* 206 (1983) 189

Gilvin, P. J., Mathieson, E. and Smith, G. C. *Nucl. Instr. Meth.* 176 (1980a) 287

Gilvin, P. J., Mathieson, E. and Smith, G. C. *IEEE Trans. Nucl. Sci.* NS-27 (1980b) 101

Gilvin, P. J., Mathieson, E. and Smith, G. C. *IEEE Trans. Nucl. Sci.* NS-28 (1981a) 835

Gilvin, P. J., Mathieson, E. and Smith, G. C. *Nucl. Instr. Meth.* 185 (1981b) 595

Golub, L., Nystrom, G., Spiller, E. and Wilczynski, J. *Proc. SPIE* 583 (1985)

Goodman, N. B. *Space. Sci. Instrum.* 2 (1976) 425

Goodrich, G. W. and Wiley, W. C. *Rev. Sci. Instrum.* 33 (1962) 761

Gordon, E. I. *IEEE Trans. Nucl. Sci.* NS-19 (1972) 190

Gordon, J. S., Mathieson, E. and Smith, G. C. *IEEE Trans. Nucl. Sci.* NS-30 (1983) 342

Gorenstein, P. and Mickiewicz, S. *Rev. Sci. Instrum.* 39 (1968) 816

Gorenstein, P. and Topka, K. *IEEE Trans. Nucl. Sci.* NS-24 (1977) 511

Gorenstein, P., Harris, B., Gursky, H. and Giacconi, R. *Nucl. Instr. Meth.* 91 (1971a) 451

Gorenstein, P., Harris, B., Gursky, H., Giacconi, R., Novick, R. and Van den Bout, P. *Science* 172 (1971b) 369

Gorenstein, P., Gursky, H., Harnden, F. R., DeCaprio, A. and Bjorkholm, P. *IEEE Trans. Nucl. Sci.* NS-22 (1975) 616

Gorenstein, P., Fabricant, D., Topka, K., Tucker, W. and Harnden, F. R. *Astrophys. J. Lett* 216 (1977) L95

Gorenstein, P., Fabricant, D., Topka, K., Harnden, F. R. and Tucker, W. *Astrophys, J.* 224 (1978) 718

Gorenstein, P., Harnden, F. R. and Fabricant, D. G. *IEEE Trans. Nucl. Sci.* NS-28 (1981) 869

Gorenstein, P., Cohen, P. and Fabricant, D. *Proc. SPIE* 597 (1985) 128

Gott, R., Parkes, W. and Pounds, K. A. *IEEE Trans. Nucl. Sci.* NS-17 (1970) 367

Gould, R. G., Judy, P. F., Klopping, J. C. and Bjarngard, B. E., *Nucl. Instr. Meth.* 144 (1977) 493

Goulding, F. S. and Landis, D. A. *IEEE Trans. Nucl. Sci.* NS-29 (1982) 1125

Gowen, R. A., Cooke, B. A., Griffiths, R. E. and Ricketts, M. J. *Mon. Not. R. Astr. Soc.* 179 (1977) 303

Greene, J. *Astrophys. J.* 130 (1959) 693

Griffiths, R. E. *Adv. Electron. Electron Phys.* 64B (1985) 483

Griffiths, R. E. *et al. IEEE Trans. Nucl. Sci.* NS-22 (1975) 611

Griffiths, R. E., Cooke, B. A., Peacock, A., Pounds, K. A. and Ricketts, M. J. *Mon. Not. R. Astr. Soc.* 175 (1976) 449

Grindlay, J. E., Garcia, M. R., Burg, R. I. and Murray, S. S. *IEEE Trans. Nucl. Sci.* NS-33 (1986) 750

Gruner, S. M., Milch, J. R. and Reynolds, G. T. *Nucl. Instr. Meth.* 195 (1982) 287

Guest, A. J. *Acta Electronica* 14 (1971) 79

Gursky, H. Naval Research Laboratory preprint NRL 41-86-016 (1986)

Gursky, H. *et al. Astrophys. J.* 146 (1966) 310

Gursky, H. *et al. Astrophys. J.* 223 (1978) 973

Hailey, C. J., Hamilton, T. T. and Ku, W. H. M. *Nucl. Instr. Meth.* 184 (1981) 543

Hailey, C. J., Ku, W. H. M. and Vartanian, M. H. *Nucl. Instr. Meth.* 213 (1983) 397

Hailey, C. J., Rockett, P., Eckart, M. and Burkhalter, P. G. *Rev. Sci. Instrum.* 56 (1985) 1553

Haller, E. E. *IEEE Trans. Nucl. Sci.* NS-29 (1982) 1109

Hanaki, H., Nagai, N., Kusakabe, T., Horiuchi, T. and Sakisaka, M. *Japan. J. Appl. Phys.* 22 (1983) 648

Hanbury Brown, R. *The Intensity Interferometer*, Taylor and Francis, London (1974)

Harnden, F. R., Fabricant, D., Topka, K., Flannery, B. P., Tucker, W. and Gorenstein, P. *Astrophys. J.* 214 (1977) 418

Harris, T. J. and Mathieson, E. *Nucl. Instr. Meth.* 96 (1971) 397

Harris, T. J. and Mathieson, E. *Nucl. Instr. Meth.* 154 (1978) 183

Harrison, D. and Kubierschky, K. *IEEE Trans. Nucl. Sci.* NS-26 (1979) 411

Harvey, P. *et al. IEEE Trans. Nucl. Sci.* NS-23 (1976) 487

Heath, R. L., Hofstadter, R. and Hughes, E. B. *Nucl. Instr. Meth.* 162 (1979) 431

Hellsing, M., Karlsson, L., Andren, H. O. and Norden, H. *J. Phys. E* 18 (1985) 920

Henke, B. *Phys. Rev. A* 2 (1972) 94

Henke, B. L., Knauer, J. P. and Premaratne, K. *J. Appl. Phys.* 52 (1981) 1509

Henkel, P., Roy, R. and Wiza J. *IEEE Trans. Nucl. Sci.* NS-25 (1978) 548

Henry, J. P., Kellogg, E. M., Briel, U. G., Murray, S. S., Van Speybroeck, L. P. and Bjorkholm, P. J. *Proc. SPIE* 106 (1977) 196

Henry, J. P., Spiller, E. and Weisskopf, M., *Proc. SPIE* 316 (1981) 166

Hettrick, M. C. and Kahn, S. M. *Proc. SPIE* 597 (1985) 291

Hill, G. E. *Adv. Electron. Electron Phys.* 40A (1976) 153

Hoan, N. N. *Nucl. Instr. Meth.* 154 (1978) 597

Hocker, L. P., Zagarino, P. A., Madrid, J., Simmons, D., Davis, B. and Lyons, P. B. *IEEE Trans. Nucl. Sci.* NS-26 (1979) 356

Holt, S. S. *Proc. Roy. Soc. A* 350 (1976) 505

Holt, S. S. *et al. Astrophys. J.* 234 (1979) L65

Holt, S. S. *et al. Proc. SPIE* 597 (1985) 267

Hopkinson, G. R. *Nucl. Instr. Meth.* 216 (1983) 423

Hopkinson, G. R. *Opt. Eng.* 26 (1987) 766

Hopkinson, G. R. and Lumb, D. H. *J. Phys. E* 15 (1982) 1214

Hughes, J. P., Long, K. S. and Novick, R. *Astrophys. J.* 280 (1984) 255

Huizenga, H., Bleeker, J. A. M., Diemer, W. H. and Huben, A. H. *Rev. Sci. Instrum.* 52 (1981) 673

Humphrey, A., Cabral, R., Brissette, R., Carroll, R., Morris, J. and Harvey, P. *IEEE Trans. Nucl. Sci.* NS-25 (1978) 445

Imhof, W. L. *et al. Astrophys. J.* 191 (1974) L7

Inoue, H., Koyama, K., Matsuoka, M., Ohashi, T., Tanaka, Y. and Tsunemi, H. *Astrophys. J.* 227 (1979) L85

Inoue, H. *et al. Nucl. Instr. Meth.* 196 (1982) 69

Ito, M., Kume, H. and Oba, K. *IEEE Trans. Nucl. Sci.* NS-31 (1984) 408

Ito, M., Yamaguchi, M. and Oba, K. *IEEE Trans. Nucl. Sci.* NS-34 (1987) 401

Iwanczyk, J. S., Dabrowski, A. J., Huth, G. C. and Drummond, W. *Adv. X-ray Analysis* 27 (1985) 405

Jacobsen, A. S. PhD thesis, Univ. California, San Diego (1968)

Jacobsen, A. S., Bishop, R. J., Culp, G. W., Jung, L., Mahoney, W. A. and Willett, J. B. *Nucl. Instr. Meth.* 127 (1975) 115

Jagoda, N., Austin, G., Mickiewicz, S. and Goddard, R. *IEEE Trans. Nucl. Sci.* NS-19 (1972) 579

Jain, A. *et al. Astron. Astrophys.* 140 (1984) 179

Jaklevic, J. M. and Goulding, F. S. *IEEE Trans. Nucl. Sci.* NS-18 (1971) 187

Janes, A. F., Pounds, K. A., Ricketts, M. J., Willmore, A. P. and Morrison, L. V. *Nature* 235 (1972) 152

Janes, A. F., Pounds, K. A., Ricketts, M. J., Willmore, A. P. and Morrison, L. V. *Nature* 244 (1973) 349

Janesick, J. R. and Blouke, M. M. *Sky and Telescope* 74 (1987) 238

Janesick, J. R., Elliott, T., McCarthy, J. K., Marsh, H. H., Collins, S. and Blouke, M. M. *IEEE Trans. Nucl. Sci.* NS-32 (1985a) 409

Janesick, J. R. *et al. Proc. SPIE* 597 (1985b) 364

Janesick, J. R., Elliott, T., Marsh, H. H., Collins, S., McCarthy, J. K. and Blouke, M. M. *Rev. Sci. Instrum.* 56 (1985c) 796

Janesick, J. R., Campbell, D., Elliott, T., Daud, T. and Ottley, P. *Proc. SPIE* 687 (1986) 36

Janesick, J. R., Elliott, T., Collins, S., Blouke, M. M. and Freeman, J. *Opt. Eng.* 26 (1987) 692

Johnstone, A. D. *et al. J. Phys. E* 20 (1987) 795

Joyce, R., Becker, R., Birsa, F., Holt, S. and Noordzy, M. *IEEE Trans. Nucl. Sci.* NS-25 (1978) 453

Kalata, K. *Nucl. Instr. Meth.* 201 (1982) 35

Kalata, K., Murray, S. S. and Chappell, J. H. *Adv. Electron. Electron Phys.* 64B (1985) 509

Kalbitzer, S. and Melzer, W. *Nucl. Instr. Meth.* 56 (1967) 301

Kayat, M. A., Rolf, D. P., Smith, G. C. and Willingale, R. *Mon. Not. R. Astr. Soc.* 191 (1980) 729

Keller, H., Klingelhofer, G. and Kankeleit, E. *Nucl. Instr. Meth. A* 258 (1987a) 221

Keller, H. *et al. J. Phys. E* 20 (1987b) 807

Kellogg, E., Henry, P., Murray, S., Van Speybroeck, L. and Bjorkholm, P. *Rev. Sci. Instrum.* 47 (1976) 282

Kellogg, E. M., Murray, S. S. and Bardas, D. *IEEE Trans. Nucl. Sci.* NS-26 (1979) 403

Kemmer, J. and Lutz, G. *Nucl. Instr. Meth. A* 253 (1987) 365

Kinzer, R. L., Johnson, W. N. and Kurfess, J. D. *Astrophys. J.* 222 (1978) 370

Kirkpatrick, P. and Baez, A. M. *J. Opt. Soc. Am.* 38 (1948) 766

Knapp, G. *Rev. Sci. Instrum.* 49 (1978) 982

Knoll, G. F. *Radiation Detection and Measurement*, Wiley, New York (1979)

Koppel, L. N. *Rev. Sci. Instrum.* 48 (1977) 669

de Korte, P. A. J., Giralt, R., Coste, J. W., Frindel, S., Flamand, J. and Contet, J. J. *Appl. Opt.* 20 (1981a) 1080

de Korte, P. A. J. *et al. Space. Sci. Rev.* 30 (1981b) 495

Kowalski, M. P., Fritz, G. G., Cruddace, R. G., Unzicker, A. E. and Swanson, N. *Appl. Opt.* 25 (1986) 2440

Kramm, I. R. and Keller, H. U. *Adv. Electron. Electron Phys.* 64A (1985) 193

Kraner, H. *IEEE Trans. Nucl. Sci.* NS-29 (1982) 1088

Kraus, H. *et al. Europhys. Lett.* 1 (1986) 161

Ku, W. H. M., Long, K., Pisarski, R. and Vartanian, M. H. IAU Symposium no. 101, *Supernova remnants and their X-ray emission*, J. Danziger and P. Gorenstein, Eds., D. Reidel, Dordrecht, (1983) p. 253

Kubierschky, K., Austin, G. K., Harrison, D. C. and Roy, A. G. *IEEE Trans. Nucl. Sci.* NS-25 (1978) 430

Kuhlmann, W. R., Lauterjung, K. H., Schimmer, B. and Sistemich, K. *Nucl. Instr. Meth.* 40 (1966) 118

Kume, H., Suzuki, S., Takeguchi, J. and Oba, K. *IEEE Trans. Nucl. Sci.* NS-32 (1985) 448

Kume, H., Maramatsu, S. and Iida, M. *IEEE Trans. Nucl. Sci.* NS-33 (1986) 359

Kurfess, J. D. and Johnson, W. N. *IEEE Trans. Nucl. Sci.* NS-22 (1975) 626

Laegsgaard, E. *Nucl. Instr. Meth.* 162 (1979) 93

Lamonds, H. A. *Nucl. Instr. Meth.* 213 (1983) 5

Lampton, M. and Carlson, C. W. *Rev. Sci. Instrum.* 50 (1979) 1093

Lampton, M. and Malina, R. F. *Rev. Sci. Instrum.* 47 (1976) 1360

Lampton, M. and Paresce, F. *Rev. Sci. Instrum.* 45 (1974) 1098

Lampton, M., Margon, B., Paresce, F., Stern, R. and Bowyer, S. *Astrophys. J. Lett.* 203 (1976a) L71

Lampton, M., Margon, B. and Bowyer, S. *Astrophys. J.* 208 (1976b) 177

Lampton, M., Siegmund, O. H. W. and Raffanti, R. *Rev. Sci. Instrum.* 58 (1987) 2298

Landecker, P. B. *IEEE Trans. Nucl. Sci.* NS-19 (1972) 463

Lapington, J. S., Schwarz, H. E. and Mason, I. M. *Proc. SPIE* 597 (1985) 222

Lapington, J. S., Smith, A. D., Walton, D. M. and Schwarz, H. E. *IEEE Trans. Nucl. Sci.* NS-34 (1987) 431

Lapson, L. B. and Timothy, J. G. *Appl. Opt.* 12 (1973) 388

Lauterjung, K. H., Pokar, I., Schimmer, B. and Staudner, R. *Nucl. Instr. Meth.* 22 (1963) 117

Lawrence, G. M. and Stone, E. J. *Rev. Sci. Instrum.* 46 (1975) 432

Leake, J. W., Smith, A., Turner, M. J. L. and White, G. *Nucl. Instr. Meth. A* 235 (1985) 589

Lecompte, P., Perez-Mendez, V. and Stoker, G. *Nucl. Instr. Meth.* 153 (1978) 543

Lemen, J. R. *et al. Solar Phys.* 80 (1982) 333

Leonov, N. V., Tyutikov, A. M. and Shishatskii, N. A. *Instr. Exp. Tech.* 22 (1980) 200

Lesser, M. P., Leach, R. W. and Angel, J. R. P. *Proc. SPIE* 627 (1986) 517

Leventhal, M., McCallum, C. J. and Watts, A. *Astrophys. J.* 216 (1977) 491

Leventhal, M., McCallum, C. J. and Stang, P. D. *Astrophys. J. Lett.* 275 (1978) L11

Levine, A., Petre, R., Rappaport, S., Smith, G. C., Evans, K. D. and Rolf, D. *Astrophys. J. Lett.* 228 (1979) L99

Ling, J. C., Mahoney, W. A., Willett, J. B. and Jacobsen, A. S. *Astrophys. J.* 231 (1979) 896

Ling, J. C., Mahoney, W. A., Wheaton, W. A. and Jacobsen, A. S. *Astrophys. J.* 275 (1983) 307

Llacer, J., Haller, E. E. and Cordi, R. C. *IEEE Trans. Nucl. Sci.* NS-24 (1977) 53

Lo, C. C. and Leskovar, B. *IEEE Trans. Nucl. Sci.* NS-28 (1981) 698

Lochner, J. C. and Boldt, E. A. *Nucl. Instr. Meth. A* 242 (1986) 382

Loty, C. *Acta Electronica* 14 (1971) 107

Lumb, D. H., Chowanietz, E. G. and Wells, A. A. *Opt. Eng.* 26 (1987) 773

Lumb, D. H. and Holland, A. D. *IEEE Trans. Nucl. Sci.* NS-35 (1988a) 534

Lumb, D. H. and Holland, A. D. *Nucl. Instr. Meth. A* 273 (1988b) 696

Lumb, D. H. and Hopkinson, G. R. *Nucl. Instr. Meth.* 216 (1983) 431

Lumb, D. H., Knight, F. K. and Schwartz, D. A. *IEEE Trans. Nucl. Sci.* NS-32 (1985) 571

McCammon, D., Moseley, S. H., Mather, J. C. and Mushotzky, R. F. *J. Appl. Phys.* 56 (1984) 1263

McCammon, D., Juda, M., Zhang, J., Kelley, R. L., Moseley, S. H. and Szymkowiak, A. E. *IEEE Trans. Nucl. Sci.* NS-33 (1986) 236

McCammon, D. *et al. Japan J. Appl. Phys.* 26 (1987), suppl. 26-3

McClintock, W. E., Barth, C. A., Steele, R. E., Lawrence, G. M. and Timothy, J. G. *Appl. Opt.* 21 (1982) 3071

McConkey, J. W., Crouch, T. and Tomc, J. *Appl. Opt.* 21 (1982) 1643

MacCuaig, N., Tajuddin, A. A., Gilboy, W. and Leake, J. W. *Nucl. Instr. Meth. A* 242 (1986) 620

McDicken, W. N. *Nucl. Instr. Meth.* 54 (1967) 157

McHardy, I. M., Lawrence, A., Pye, J. P. and Pounds, K. A. *Mon. Not. R. Astr. Soc.* 197 (1982) 893

Mackay, C. D. *Ann. Rev. Astron. Astrophys.* 24 (1986) 255

McKee, B. T. A., Stewart, A. T. and Vesel, J. *Nucl. Instr. Meth. A* 234 (1985) 191

Mackenzie, J. D. MCP glass analysis studies, final technical report, Materials department, UCLA (1977)

McKenzie, J. M. *Nucl. Instr. Meth.* 162 (1979) 49

McMullan, W. G., Charbonneau, S. and Thewalt, M. L. W. *Rev. Sci. Instrum.* 58 (1987) 1626

Madden, N. W., Hanepen, G. and Clark, B. C. *IEEE Trans. Nucl. Sci.* NS-33 (1986) 303

Mahoney, W. A., Ling, J. C., Jacobsen, A. S. and Tapphorn, R. M. *Nucl. Instr. Meth.* 178 (1980) 363

Mahoney, W. A., Ling, J. C. and Jacobsen, A. S. *Nucl. Instr. Meth.* 185 (1981) 449

Mahoney, W. A., Ling, J. C., Wheaton, W. A. and Jacobsen, A. S. *Astrophys. J.* 286 (1984) 598

Makino, F., and the Astro-C team. *Astrophys. Lett.* 25 (1987) 223

Malina, R. F. and Coburn, K. R. *IEEE Trans. Nucl. Sci.* NS-31 (1984) 404

Malina, R. F., Jelinsky, P. and Bowyer, S. *Proc. SPIE* 597 (1985) 154

Malm, H. L. *IEEE Trans. Nucl. Sci.* NS-19 (1972) 263

Manzo, G., Peacock, A., Andresen, R. D. and Taylor, B. G. *Nucl. Instr. Meth.* 174 (1980) 301

Manzo, G., Re, S., Gerardi, G., Fici, M., Giarrusso, S. and Mistretta, O. *IEEE Trans. Nucl. Sci.* NS-32 (1985) 150

Maor, D. and Rosner, B. *J. Phys. E* 11 (1978) 1141

Margon, B. and Bowyer, S. *Sky and Telescope* 50 (1975) 4

Marshall, N., Warwick, R. S. and Pounds, K. A. *Mon. Not. R. Astr. Soc.* 194 (1981) 987

Martin, B. R. *Statistics for Physicists*, Academic Press, London (1971)

Martin, C., Jelinsky, P., Lampton, M., Malina, R. F. and Anger, H. O. *Rev. Sci. Instrum.* 52 (1981) 1067

Martin, C. and Bowyer, S. *Appl. Opt.* 21 (1982) 4206

Mason, I. M. and Culhane, J. L. *IEEE Trans. Nucl. Sci.* NS-30 (1983) 485

Mason, I. M., Branduardi-Raymont, G., Culhane, J. L., Corbet, R. H. D., Ives, J. C. and Sanford, P. W. *IEEE Trans. Nucl. Sci.* NS-31 (1984) 795

Massey, H. and Robins, M. O. *The History of British Space Science*, Cambridge University Press (1986)

Mathieson, E. *J. Phys. E* 2 (1979) 183

Mathieson, E. and Charles, M. W. *Nucl. Instr. Meth.* 72 (1969) 155

Mathieson, E. and Harris, T. J. *Nucl. Instr. Meth.* 154 (1978) 189

Mathieson, E. and Sanford, P. W. *Proc. Intern. Symp. Nucl. Elect. (ENEA)* (1963a) 65

Mathieson, E. and Sanford, P. W. *J. Sci. Instrum.* 40 (1963b) 446

Mathieson, E., Smith, G. C. and Gilvin, P. J. *Nucl. Instr. Meth.* 174 (1980) 221

Matoba, M., Hirose, T., Sakae, T., Kametani, H., Ijiri, H. and Shintaki, T. *IEEE Trans. Nucl. Sci.* NS-32 (1985) 541

Matsura, S. *et al. IEEE Trans. Nucl. Sci.* NS-32 (1985) 350

Matteson, J., Nolan, P., Paciesas, W. S. and Pelling, R. M. University of California, San Diego, preprint no. UCSD SP76-07 (1976)

Mels, W. A., Lowes, P., Brinkman, A. C., Naber, A. P. and Rook, A. *Nucl. Instr. Meth. A* 273 (1988) 689

Menefee, J., Cho, Y. and Swinehart, C. *IEEE Trans. Nucl. Sci.* NS-14 (1967) 464

Mertz, L. and Young, N. O. *Proc. Int. Conf. Opt. Instrum.*, London (1961) 305

Metzner, G. *Nucl. Instr. Meth. A* 242 (1986) 493

Mewe, R. and Gronenschild, E. H. B. M. *Astron. Astrophys. Suppl.* 45 (1981) 11

Mewe, R. *et al. Astrophys. J.* 260 (1982) 233

Meyerott, A. J., Fisher, P. C. and Roethig, D. T. *Rev. Sci. Instrum.* 35 (1964) 669

Michau, J. C., Pichard, B., Soirat, J. P., Spiro, M. and Vignaud, D. *Nucl. Instr. Meth. A* 244 (1986) 565

Michette, A. G. *J. Photograph. Sci.* 32 (1984) 45

Miller, G. L. *IEEE Trans. Nucl. Sci.* NS-19 (1972) 251

Morenzoni, E., Oba, K., Pedroni, E. and Taqqu, D. *Nucl. Instr. Meth. A* 263 (1988) 397

Morgan, R. *J. Phys. E* 4 (1971) 372

Mortara, L. and Fowler, A. *Proc. SPIE* 290 (1981) 28

Morton, J. M. and Parkes, W. *Acta Electronica* 16 (1973) 85

Moseley, S. H., Mather, J. C. and McCammon, D. *J. Appl. Phys.* 56 (1984) 1257

Moseley, S. H., Kelley, R. L., Mather, J. C., Mushotzky, R. F., Szymkowiak,

A. E. and McCammon, D. *IEEE Trans. Nucl. Sci.* NS-32 (1985) 134

Moszynski, M., Vacher, J. and Odru, R. *Nucl. Instr. Meth.* 217 (1983) 453

Mullard Technical Information, M81-0151, Mullard Ltd., London (1981)

Murray, S. S. and Chappell, J. H. *Proc. SPIE* 597 (1985) 274

Murray, S. S., Fabbiano, G., Fabian, A. C., Epstein, A. and Giacconi, R. *Astrophys. J. Lett.* 234 (1979) L69

Musket, R. G. *Nucl. Instr. Meth.* 117 (1974) 385

Naday, I., Strauss, M. G., Sherman, I. S., Kramer, M. R. and Westbrook E. M. *Opt. Eng.* 26 (1987) 788

Nakano, G. H., Imhof, W. L., Reagan, J. B. and Johnson, R. G. *IEEE Trans. Nucl. Sci.* NS-21 (1974) 159

Nakano, G. H., Imhof, W. L. and Reagan, J. B. *IEEE Trans. Nucl. Sci.* NS-27 (1980) 405

Narayan, R. and Nityananda, R. *Ann. Rev. Astron. Astrophys.* 24 (1986) 127

Nestor, O. H. and Huang, C. Y. *IEEE Trans. Nucl. Sci.* NS-22 (1975) 68

Niinikoski, T. O. and Udo, F. CERN internal report 74-6 (1974)

Nilsson, O., Hasselgren, L., Siegbahn, K., Berg, S., Andersson, L. P. and Tove, P. A. *Nucl. Instr. Meth.* 84 (1970) 301

Nousek, J. A., Garmire, G. P., Ricker, G. R., Collins, S. A. and Riegler, G. R. *Astro. Lett. Comm.* 26 (1987) 35

Novick, R. In *Planets, Stars and Nebulae Studied with Photo-polarimetery*, T. Gehrels, ed. University of Arizona Press (1974) p. 262

Novick, R., Weisskopf, M. C., Berthelsdorf, R., Linke, R. and Wolff, R. S. *Astrophys. J. Lett.* 174 (1972) L1

Novick, R., Weisskopf, M. C., Silver, E. M., Kestenbaum, H. L., Long, K. S. and Wolff, R. S. *Ann. NY Acad. Sci.* 302 (1977) 312

Novick, R., Chanan, G. and Helfand, D. J. *Proc. ESA workshop 'Cosmic X-ray spctroscopy mission'*, ESA SP-239 (1985) 265

Oba, K. and Rehak, P. *IEEE Trans. Nucl. Sci.* NS-28 (1981) 683

Oba, K., Sugiyama, M., Suzuki, Y. and Yoshimura,Y. *IEEE Trans. Nucl. Sci.* NS-26 (1979) 346

Oba, K., Rehak, P. and Smith, S. D. *IEEE Trans. Nucl. Sci.* NS-28 (1981) 705

Oda, M. *Appl. Opt.* 4 (1965) 143

Ogawara, Y. *et al. Nature* 295 (1982) 675

Ortale, C., Padgett, L. and Schnepple, W. F. *Nucl. Instr. Meth.* 213 (1983) 95

Oshchepkov, P. K., Skvortsov, B. N., Osanov, B. A. and Siprikov, I. V. *Instr. Exp. Tech.* 4 (1960) 611

Overley, J. C., Lefevre, H. W., Nolt, I. G., Radostitz, J. V., Predko, S. and Ade, P. A. R. *Nucl. Instr. Meth. B* 10/11 (1985) 928

Owens, A. *Nucl. Instr. Meth. A* 238 (1985) 473

Paciesas, W. *et al. Nucl. Instr. Meth.* 215 (1983) 261

Panitz, J. A. and Foesch, J. A. *Rev. Sci. Instrum.* 47 (1976) 44

Paresce, F. *Appl. Opt.* 14 (1975) 2823

Parkes, W. and Gott, R. *Nucl. Instr. Meth.* 95 (1971) 487

Parkes, W., Gott, R. and Pounds, K. A. *IEEE Trans. Nucl. Sci.* NS-17 (1970) 360

Parkes, W., Evans, K. D. and Mathieson, E. *Nucl. Instr. Meth.* 121 (1974) 151

Parmar, A. N. and Smith, A. *EXOSAT Express* 10, April (1985)

Peacock, A., Andresen, R. D., Leimann, E. A., Long, A., Manzo, G. and Taylor, B. G. *Nucl. Instr. Meth.* 169 (1980) 613

Peacock, A. *et al. Space Sci. Rev.* 30 (1981) 525

Peacock, A., Taylor, B. G., White, N., Courvoisier, T. and Manzo, G. *IEEE Trans. Nucl. Sci.* NS-32 (1985) 108

Pearson, J. F., Fraser, G. W. and Whiteley, M. J. *Nucl. Instr. Meth. A* 258 (1987) 270

Pearson, J. F., Lees, J. E. and Fraser, G. W. *IEEE Trans. Nucl. Sci.* NS-35 (1988) 520

Peckerar, M. C., Baker, W. D. and Nagel, D. J. *J. Appl. Phys.* 48 (1977) 2565

Peckerar, M. C., McCann, D. H. and Yu, L. *Appl. Phys. Lett.* 39 (1981) 55

Pehl, R. H. and Goulding, F. S. *Nucl. Instr. Meth.* 81 (1970) 329

Pehl, R. H., Goulding F. S., Landis, D. A. and Lenslinger, M. *Nucl. Instr. Meth.* 58 (1968) 45

Pell, E. M. *J. Appl. Phys.* 31 (1960) 291

Penning, F. M. *Physica* 1 (1934) 1028

Peterson, L. E. *Ann. Rev. Astron. Astrophys.* 13 (1975) 423

Peterson, L. E., Pelling, R. M. and Matteson, J. L. *Space Sci. Rev.* 13 (1972) 320

Petre, R. and Serlemitsos, P. *J. Appl. Opt.* 24 (1985) 1833

Pfeffermann, E. and Briel, U. G. *Proc. SPIE* 597 (1985) 208

Pfeffermann, E., Aschenbach, B., Brauninger, H. and Trümper, J. *Space Sci. Rev.* 30 (1981) 251

Pfeffermann, E. *et al. Proc. SPIE* 733 (1987) 519

Polaert, R. and Rodiere, J. *Phillips Tech. Rev.* 34 (1974) 270

Policarpo, A. J. P. L. *Nucl. Instr. Meth.* 153 (1978) 389

Policarpo, A. J. P. L., Alves, M. A. F., Dos Santos, M. C. M. and Carvalho, M. J. T. *Nucl. Instr. Meth.* 102 (1972) 337

Policarpo, A. J. P. L., Alves, M. A. F., Salete, M., Leite, S. C. P. and Dos Santos, M. C. M. *Nucl. Instr. Meth.* 118 (1974) 221

Ponman, T. *Mon. Not. R. Astr. Soc.* 196 (1981) 583

Ponman, T. *Nucl. Instr. Meth.* 221 (1984) 72

Pounds, K. A. *Nature Phys. Sci.* 229 (1971) 175

Pounds, K. A. *Q. J. Roy. Astr. Soc.* 27 (1986) 435

Premaratne, K., Dietz, E. R. and Henke, B. L. *Nucl. Instr. Meth.* 207 (1983) 465

Primini, F. A. *et al. Astrophys. J.* 243 (1981) L13

Proctor, R. J., Skinner, G. K. and Willmore, A. P. *Mon. Not. R. Astr. Soc.* 185 (1978) 745

Proctor, R. J., Skinner, G. K. and Willmore, A. P. *Mon. Not. R. Astr. Soc.* 187 (1979) 633

Proctor, R., Pietsch, W. and Reppin C., *Nucl. Instr. Meth.* 197 (1982) 477

Purshke, M., Nuxoll, W., Gaul, G. and Santo, R. *Nucl. Instr. Meth. A* 261 (1987) 537

Radeka, V. and Boie, R. A. *Nucl. Instr. Meth.* 178 (1980) 543

Rager, J. P. and Renaud, J. F. *Rev. Sci. Instrum.* 45 (1974) 922

Rager, J. P., Renaud, J. F. and Tezenas du Montcel, V. *Rev. Sci. Instrum.* 45 (1974) 927

Ramsey, B. Weisskopf, M. C. and Elsner, R. F. *Proc. SPIE* 597 (1985) 213

Ramsey, B. D. and Weisskopf, M. C. *Nucl. Instr. Meth. A* 248 (1986) 550

Rapley, C. G., Culhane, J. L., Acton, L. W., Catura, R. C., Joki, E. G. and Bakke, J. C. *Rev. Sci. Instrum.* 48 (1977) 1123

Rappaport, S., Petre, R., Kayat, M. A., Evans, K. D., Smith, G. C. and Levine, A. *Astrophys. J.* 227 (1979) 285

Rawlings, K. J. *Nucl. Instr. Meth. A* 253 (1986) 85

Raymond, J. C. and Smith, B. W. *Astrophys. J. Suppl.* 35 (1977) 419

Read, P. D., Powell, J. R., van Breda, I. G., Lyons, A. and Ridley, N. R. *Proc. SPIE* 627 (1986) 645

Rees, D., McWhirter, I., Rounce, P. A., Barlow, F. E. and Kellock, S. J. *J. Phys. E* 13 (1980) 763

Rehak, P. *et al. Nucl. Instr. Meth. A* 248 (1986) 367

Reme, H. *et al. J. Phys. E* 20 (1987) 721

Richter, L. J. and Ho, W. *Rev. Sci. Instrum.* 57 (1986) 1469

Ricker, G. R., Vallerga, J. V., Dabrowski, A. J., Iwanczyk, J. S. and Entine, G. *Rev. Sci. Instrum.* 53 (1982) 700

Ricker, G. R., Vallerga, J. V. and Wood, D. R. *Nucl. Instr. Meth.* 213 (1983) 133

Riegler, G. R. and Moore, K. A. *IEEE Trans. Nucl. Sci.* NS-20 (1973) 102

Riegler, G. R., Garmire, G., Moore, W. E. and Stevens, J. C. *Bull. Am. Phys. Soc.* 15 (1970) 635

Riegler, G. R., Ling, J. C., Mahoney, W. A., Wheaton, W. A. and Jacobsen, A. S. *Astrophys. J. Lett.* 294 (1985) L13

Roberts, E., Stapinski, T. and Rodgers, A. *J. Opt. Soc. Am. A* 3 (1986) 2146

Rocchia, R. *et al. Astron. Astrphys.* 130 (1984) 53

Rogers, D. and Malina, R. F. *Rev. Sci. Instrum.* 53 (1982) 1438

Rossi, B. and Staub, H. H. *Ionisation Chambers and Counters – Experimental Techniques*, McGraw-Hill, New York (1949)

Rothschild, R. E., Baity, W. A., Marscher, A. P. and Wheaton, W. A. *Astrophys. J. Lett.* 243 (1981) L9

Rougeot, H., Roziere, G. and Driard, B. *Adv. Electron. Electron Phys.* 52 (1979) 227

Ruggieri, D. J. *IEEE Trans. Nucl. Sci.* NS-19 (1972) 74

Russell, P. C. and Pounds, K. A. *Nature* 209 (1966) L490

Rutherford, E. and Geiger, H. *Proc. Roy. Soc. A* 81 (1908) 141

Saloman, E. B., Pearlman, J. S. and Henke, B. L. *Appl. Opt.* 19 (1980) 749

Sams, B. J., Golub, L. and Kalata, K. Harvard-Smithsonian Center for Astrophysics preprint no. 2487 (1987)

Samson, J. A. R. *Techniques of Vacuum Ultraviolet Spectroscopy*, Wiley, New York (1967)

Sandage, A. R. *et al. Astrophys. J.* 146 (1966) 316

Sandel, B. R., Broadfoot, A. L. and Shemansky, D. E. *Appl. Opt.*
16 (1977) 1435

Sanford, P. W. and Ives, J. C. *Proc. Roy. Soc. A* 350 (1976) 491

Sanford, P. W., Cruise, A. M. and Culhane, J. L. In *Non-Solar X- and Gamma-ray Astronomy*, L. Gratton, ed., IAU Symp. no. 37, D. Reidel, Dordrecht (1970), p. 35

Sanford, P. W., Mason, I., Dimmock, K. and Ives, J. C. *IEEE Trans. Nucl. Sci.* NS-26 (1979) 169

Schagen, P. *Phil. Trans. Roy. Soc. A* 269 (1971) 233

Schieber, M. *Nucl. Instr. Meth.* 144 (1977) 469

Schieber, M., ed. 'Proc. 1st int. workshop on HgI_2/CdTe nuclear radiation detectors', *Nucl. Instr. Meth.* 150 (1978)

Schieber, M. and van den Berg, L., eds. 'Proc. 5th int. workshop on HgI_2 nuclear radiation detectors', *Nucl. Instr. Meth.* 213 (1983)

Schnopper, H. W. and Thompson, R. I. *Space Sci. Rev.* 8 (1968) 534

Schnopper, H. W. *et al. Astrophys. J.* 253 (1982) 131

Schwartz, D. A., Griffiths, R. E., Murray, S. S., Zombeck, M. V., Barrett, J. and Bradley, W. *Proc. SPIE* 184 (1979) 247

Schwartz, D. A., Knight, F. K., Lumb, D. H., Chowanietz, E. G. and Wells, A. A. Proc. *SPIE Symp. West*, San Diego (1985)

Schwarz, H. E. *Nucl. Instr. Meth. A* 238 (1985) 124

Schwarz, H. E. and Mason, I. M. *Nature* 309 (1984) 532

Schwarz, H. E. and Mason, I. M. *IEEE Trans. Nucl. Sci.* NS-32 (1985) 516

Schwarz, H. E., Thornton, J. and Mason, I. M. *Nucl. Instr. Meth.* 225 (1984) 325

Schwarz, H. E., Lapington, J. S. and Culhane, J. L. *Proc. SPIE* 597 (1985a) 397

Schwarz, H. E., Lapington, J. S., Hilliard, H. S., Rose, J. G., Sheather, P. H. and Culhane, J. L. *Proc. SPIE* 597 (1985b) 178

Sephton, J. P., Turner, M. J. L. and Leake, J. W. *Nucl. Instr. Meth.* 219 (1984) 534

Seidel, W., Oberauer, L. and v. Feilitzsch, F. *Rev. Sci. Instrum.* 58 (1987) 1471

Serlemitsos, P., Petre, R., Glasser, C. and Birsa, F. *IEEE Trans. Nucl. Sci.* NS-31 (1984) 786

Seward, F. *et al. Appl. Opt.* 21 (1982) 2012

Shalev, S. and Hopstone, P. *Nucl. Instr. Meth.* 155 (1978) 237

Short, M. *Nucl. Instr. Meth.* 221 (1984) 142

Siddiqui, S. H. *J. Appl. Phys.* 48 (1977) 3053

Siddiqui, S. H. *IEEE Trans. Electron Devices* ED-26 (1979) 1059

Siegmund, O. H. W., Culhane, J. L., Mason, I. M. and Sanford, P. W. *Nature* 295 (1982) 678

Siegmund, O. H. W., Clothier, S., Culhane, J. L. and Mason, I. M. *IEEE Trans. Nucl. Sci.* NS-30 (1983a) 350

Siegmund, O. H. W., Culhane, J. L. and Mason, I. M. *Adv. Space Res.* 2 (1983b) 229

Siegmund, O. H. W., Coburn, K. and Malina, R. F. *IEEE Trans. Nucl. Sci.* NS-32 (1985) 443

Siegmund, O. H. W., Lampton, M., Bixler, J., Bowyer, S. and Malina, R. F. *IEEE Trans. Nucl. Sci.* NS-33 (1986a) 724

Siegmund, O. H. W., Everman, E., Vallerga, J. V., Labov, S., Bixler, J. and Lampton, M. *Proc. SPIE* 687 (1986b) 117

Siegmund, O. H. W., Vallerga, J. and Jelinsky, P. *Proc. SPIE* 689 (1986c) 40

Siegmund, O. H. W., Everman, E., Vallerga, J. V., Sokolowski, J. and Lampton, M. *Appl. Opt.* 26 (1987) 3607

Siegmund, O. H. W., Everman, E., Vallerga, J. and Lampton, M. *Proc. SPIE/ANRT* 209 (1988a) (in press)

Siegmund, O. H. W., Vallerga, J. and Warglin, B. *IEEE Trans. Nucl. Sci.* NS-35 (1988b) 524

Simons, D. G., de Korte, P. A. J., Peacock, A. and Bleeker, J. A. M. *Proc. SPIE* 597 (1985a) 190

Simons, D. G., de Korte, P. A. J., Peacock, A., Smith, A. and Bleeker, J. A. M. *IEEE Trans. Nucl. Sci.* NS-32 (1985b) 345

Simons, D. G., Fraser, G. W., de Korte, P. A. J., Pearson, J. F. and de Jong, L. *Nucl. Instr. Meth. A* 261 (1987) 579

Simons, D. G., de Korte, P. A. J. and Heppener, M. *Nucl. Instr. Meth. A* 273 (1988) 512

Sims, M. R., Peacock, A. and Taylor, B. G. *IEEE Trans. Nucl. Sci.* NS-30 (1983) 394

Sims, M. R., Peacock, A. and Taylor, B. G. *Nucl. Instr. Meth.* 221 (1984) 168

Sims, M. R. *et al. Astrophys. Space Sci.* 116 (1985) 61

Singer, S., Aiello, W. P., Bergey, J. A., Edeskuty, F. J. and Williamson, K. D. *IEEE Trans. Nucl. Sci.* NS-19 (1972) 626

Sipila, H. *Nucl. Instr. Meth.* 133 (1976) 251

Sipila, H. *IEEE Trans. Nucl. Sci.* NS-26 (1979) 181

Skinner, G. K. *Nucl. Instr. Meth.* 221 (1984) 33

Skinner, G. K. *et al. Nature* 330 (1987) 544

Smith, A. 'The Soviet space astronomy programme', ESLAB preprint 87/176 (1987)

Smith, A. and Turner, M. J. L. *Nucl. Instr. Meth.* 192 (1982) 475

Smith, A., Jones, L. R., Peacock, A. and Pye, J. P. *Astrophys. J.* 296 (1985a) 469

Smith, A., Peacock, A. and Kowalski, T. Z. *IEEE Trans. Nucl. Sci.* NS-34 (1987) 57

Smith, A. D. and Allington-Smith, J. R. *IEEE Trans. Nucl. Sci.* NS-33 (1986) 295

Smith, D. G. and Pounds, K. A. *IEEE Trans. Nucl. Sci.* NS-15 (1968) 541

Smith, G. C., Pearson, J. F. and Mathieson, E. *Nucl. Instr. Meth.* 192 (1982) 383

Smith, G. C., Fischer, J. and Radeka, V. *IEEE Trans. Nucl. Sci.* NS-31 (1984) 111

Smith, G. C., Fischer, J. and Radeka, V. *IEEE Trans. Nucl. Sci.* NS-32 (1985b) 521

Smith, J. F. and Courtier, G. M. *Proc. Roy. Soc. A* 350 (1976) 421

Smith, P. F. 'Possible experiments for direct detection of particle candidates for the galactic dark matter', Rutherford Appleton Laboratory report RAL-86-029 (1986)

Sobottka, S. E. and Williams, M. B. *IEEE Trans. Nucl. Sci.* NS-35 (1988) 348

Somer, T. A. and Graves, P. W. *IEEE Trans. Nucl. Sci.* NS-16 (1969) 376

Somner, A. H. *Photoemissive Materials*, Wiley, New York (1968)

Soul, P. B. *Nucl. Instr. Meth.* 97 (1971) 555

Spindt, C. A. and Shoulders, K. R. *Rev. Sci. Instrum.* 36 (1965) 775

Squillante, M. R., Farrell, R., Lund, J. C., Sinclair, F., Entine, G. and Keller, K. R. *IEEE Trans. Nucl. Sci.* NS-33 (1986) 336

Stair, P. C. *Rev. Sci. Instrum.* 51 (1980) 132

Stephan, K. H. and Englhauser, J. *Proc. SPIE* 689 (1986) 128

Stern, R. A., Liewer, K. and Janesick, J. R. *Rev. Sci. Instrum.* 54 (1983) 198

Stern, R. A., Catura, R. C., Blouke, M. M. and Winzenread, M. *Proc. SPIE* 627 (1986) 583

Stern, R. A. *et al. Opt. Eng.* 26 (1987) 875

Stevens, J. C. and Garmire, G. P. *Astrophys. J. Lett.* 180 (1973) L19

Stone, R. E., Barkley, V. A. and Fleming, J. A. *IEEE Trans. Nucl. Sci.* NS-33 (1986) 299

Stroke, H. H. *et al. IEEE Trans. Nucl. Sci.* NS-33 (1986) 759

Stumpel, J. W., Sanford, P. W. and Goddard, H. F. *J. Phys. E* 6 (1973) 397

Sumner, T. J., Rochester, G. K. and Hall, G. *Nucl. Instr. Meth. A* 273 (1988) 701

Sunyaev, R. *et al. Nature* 330 (1987) 227

Sutherland, P. G., Weisskopf, M. C. and Kahn, S. M. *Astrophys. J.* 219 (1978) 1029

Tadjine, A., Gosselin, D., Koebel, J. M. and Siffert, P. *Nucl. Instr. Meth.* 213 (1983) 77

Tagaki, S. *et al. Nucl. Instr. Meth.* 215 (1983) 207

Tananbaum, H. and Kellogg, E. M. *IEEE Trans. Nucl. Sci.* NS-17 (1970) 147

Tananbaum, H. *et al. Astrophys. J. Lett.* 234 (1979) L9

Taylor, B. G. *EXOSAT Express* 11 (1985) 23

Taylor, B. G., Andresen, R. D., Peacock, A. and Zobl, R. *Space Sci. Rev.* 30 (1981) 479

Taylor, B. G., Sims, M. R., Davelaar, J. and Peacock, A. *Adv. Space Res.* 2 (1983a) 225

Taylor, R. C., Hettrick, M. C. and Malina, R. F. *Rev. Sci. Instrum.* 54 (1983b) 171

Terell, J., Priedhorsky, W. C., Belian, R. D., Conner, J. P. and Evans, W. D. *Ann. NY Acad. Sci.* 422 (1984) 282

Thewlis, J., ed. *Encyclopaedic Dictionary of Physics*, Pergamon, Oxford (1962)

Thomas, H. D. and Turner, M. J. L. *IEEE Trans. Nucl. Sci.* NS-30 (1983) 499

Thomas, H. D. and Turner, M. J. L. *Nucl. Instr. Meth.* 221 (1984) 175

Thomson, J. J. *Phil. Mag.* 47 (1899) 253

Thorne, D. J., Jorden, P. R., Waltham, N. R. and van Breda, I. G. *Proc. SPIE* 627 (1986) 530

Timothy, A. F. and Timothy, J. G. *J. Phys. E* 2 (1969) 825

Timothy, J. G., *Rev. Sci. Instrum.* 45 (1974) 834

Timothy, J. G. *Rev. Sci. Instrum.* 52 (1981) 1131

Timothy, J. G. *Proc. SPIE* 597 (1985) 330

Timothy, J. G. *Proc. SPIE* 691 (1986) 35

Timothy, J. G. Paper presented at London position sensitive detector conference, Sept. 1987

Timothy, J. G. and Bybee, R. L. *Appl. Opt.* 14 (1975) 1632

Timothy, J. G. and Bybee, R. L. *Rev. Sci. Instrum.* 48 (1977) 292

Timothy, J. G. and Bybee, R. L. *Rev. Sci. Instrum.* 50 (1979) 743

Timothy, J. G. and Bybee, R. L. *Proc. SPIE* 687 (1986) 109

Timothy, J. G., Graves, P. W., Loretz, T. J. and Roy, R. L. *Proc. SPIE* 687 (1986) 104

Tomc, J., Zetner, P., Westerveld, W. B. and McConkey, J. W. *Appl. Opt.* 23 (1984) 656

Tomitani, T. *Nucl. Instr. Meth.* 100 (1972) 179

Torr, M. R., Torr, D. G., Baum, R. and Spielmaker, R. *Appl. Opt.* 25 (1986) 2768

Trap, H. J. L. *Acta Electronica* 14 (1971) 41

Trümper, J. *Physica Scripta* T7 (1984) 209

Trümper, J., Pietsch, W., Reppin, C., Voges, W., Staubert, R. and Kendziorra, E. *Astrophys. J. Lett.* 219 (1978) L105

Tucker, W. and Giacconi, R. *The X-ray Universe*, Harvard University Press (1985)

Turner, M. J. L. and Breedon, L. M. *Mon. Not. R. Astr. Soc.* 208 (1984) 29p

Turner, M. J. L., Smith, A. and Zimmermann, H. U. *Space Sci. Rev.* 30 (1981) 513

Twerenbold, D. *Europhys. Lett.* 1 (1986) 209

Twerenbold, D. *Nucl. Instr. Meth. A* 260 (1987) 430

Twerenbold, D. *Nucl. Instr. Meth. A* 273 (1988) 575

Ubertini, P. *et al. Nucl. Instr. Meth.* 217 (1983) 97

Underwood, J. H., Barbee, T. W. and Keith, D. C. *Proc. SPIE* 184 (1979) 123

Vaiana, G. S., Reidy, W. P., Zehnpfennig, T., Van Speybroeck, L. P. and Giacconi, R. *Science* 161 (1968) 564

Vaiana, G. S. *et al. Astrophys. J. Lett.* 185 (1973) L47

Vaiana, G. S. *et al. Astrophys. J.* 245 (1981) 163

Valette, C. and Waysand, G. In *New Instrumentation for Astronomy*, K. A. Van der Hucht and G. Vaiana, eds., Pergamon Press, Oxford (1977), p. 313

Vallerga, J. V. and Lampton, M. *Proc. SPIE* 868 (1988) (in press)

Vallerga, J., Ricker, G. R., Schnepple, W. S. and Ortale, C. *IEEE Trans. Nucl. Sci.* NS-29 (1982) 151

Vallerga, J. V., Vanderspek, R. K. and Ricker, G. R. *Nucl. Instr. Meth.* 213 (1983) 145

Vallerga, J. V., Siegmund, O. H. W., Everman, E. and Jelinsky, P. *Proc. SPIE* 689 (1986) 138

van der Klis, M. *et al. Nature* 316 (1985) 225

Van Hoof, H. A. and Van der Wiel, M. J. *J. Phys. E* 13 (1980) 409

Van Lint, V. A. J. *Nucl. Instr. Meth. A* 253 (1987) 453

Van Speybroeck, L., Kellogg, E., Murray, S. S. and Duckett, S. *IEEE Trans. Nucl. Sci.* NS-21 (1974) 408

Varnell, L. S. *et al. IEEE Trans. Nucl. Sci.* NS-31 (1984) 300

Vartanian, M. H., Lum, K. S. K. and Ku, W. H. M. *Astrophys. J. Lett.* 288 (1985) L5

Va'vra, J. *Nucl. Instr. Meth. A* 252 (1986) 547

Villa, G. *et al. Mon. Not. R. Astr. Soc.* 176 (1976) 609

von Ammon, W. and Herzer, H. *Nucl. Instr. Meth. A* 226 (1984) 94

Walton, D. M., Stern, R. A., Catura, R. C. and Culhane, J. L. *Proc. SPIE* 501 (1984)

Walton, D. M., Culhane, J. L., Stern, R. A. and Catura, R. C. *Proc. ESA workshop, 'Cosmic X-ray spectroscopy mission'*, ESA SP-239 (1985) 291

Wang, C. L., Leipelt, G. R. and Nilson, D. G. *Rev. Sci. Instrum.* 56 (1985) 833

Warwick, R. S. *et al. Mon. Not. R. Astr. Soc.* 197 (1982) 865

Washington, D. *Nucl. Instr. Meth.* 111 (1973) 573

Washington, D., Duchenois, V., Polaert, R. and Beasley, R. M. *Acta Electronica* 14 (1971) 201

Watson, M. G., Willingale, R., Grindlay, J. E. and Hertz, P. *Astrophys. J.* 250 (1981a) 142

Watson, M. G., Willingale, R., Grindlay, J. E. and Hertz, P. *Space Sci. Rev.* 30 (1981b) 293

Watson, M. G., King, A. R. and Osborne, J. *Mon. Not. R. Astr. Soc.* 212 (1985) 917

Watson, M. G., Stewart, G. C., Brinkmann, W. and King, A. R. *Mon. Not. R. Astr. Soc.* 222 (1986) 261

Weast, R. C., ed. *Handbook of Chemistry and Physics*, 65th ed, CRC Press, Boca Raton, FL (1985)

Webb, P. P. and McIntyre, R. J. *IEEE Trans. Nucl. Sci.* NS-23 (1976) 138

Webb, S. *et al. Nucl. Instr. Meth.* 221 (1984) 233

Weiser, H., Vitz, R. C., Moos, H. W. and Weinstein, A. *Appl Opt.* 15 (1976) 3123

Weissenberger, E., Kast, W. and Gonnenwein, F. *Nucl. Instr. Meth.* 163 (1979) 357

Weisskopf, M. *Proc. SPIE* 597 (1985) 228

Weisskopf, M. C., Sutherland, P. G., Elsner, R. F. and Ramsey, B. D., *Nucl. Instr. Meth. A* 236 (1985) 371

Wells, A. *Proc. SPIE* 597 (1985) 146

Wells, A., Lumb, D. H. and Willingale, R. *Proc. ESA workshop 'Cosmic X-ray spectroscopy mission'*, ESA SP-239 (1985) 297

Whiteley, M. J., Pearson, J. F., Fraser, G. W. and Barstow, M. A. *Nucl. Instr. Meth.* 224 (1984) 287

Wickersheim, K. A., Alves, R. V. and Buchanan, R. A. *IEEE Trans. Nucl. Sci.* NS-17 (1970) 57

Wijnaendts van Resandt, R. W., den Harink, H. C. and Los, J. *J. Phys. E* 9 (1976) 503

Wiley, W. C. and Hendee, C. F. *IRE Trans. Nucl. Sci.* NS-9 (1962) 103

Wilken, B., Weiss, W., Studemann, W. and Haseber, N. *J. Phys. E* 20 (1987) 778

Wilkinson, D. H. *Ionisation Chambers and Counters*, Cambridge University Press (1950)

Willingale, R. *Mon. Not. R. Astr. Soc.* 194 (1981) 359

Willingale, R. *Nucl. Instr. Meth.* 221 (1984) 1

Willingale, R., Sims, M. R. and Turner, M. J. L. *Nucl. Instr. Meth.* 221 (1984) 60

Willmore, A. P., Skinner, G. K., Eyles, C. J. and Ramsey, B. *Nucl. Instr. Meth.* 221 (1984) 284

Wiza, J., Henkel, P. and Roy, R. *Rev. Sci. Instrum.* 48 (1977) 1217

Wiza, J. L. *Nucl. Instr. Meth.* 162 (1979) 587

Wojcik, M. and Grotowski, K. *Nucl. Instr. Meth.* 178 (1980) 189

Wolter, H. *Ann. Physik* 10 (1952a) 94

Wolter, H., *Ann. Physik* 10 (1952b) 286

Woodhead, A. W. and Ward, R. *Radio Electron. Eng.* 47 (1977) 545

Wyman, C. L. *et al. Proc. SPIE* 597 (1985) 2

Yee, J. H., Sherohman, J. W. and Armantrout, G. A. *IEEE Trans. Nucl. Sci.* NS-23 (1976) 117

Zarnecki, J. C. and Culhane, J. L. *Mon. Not. R. Astr. Soc.* 178 (1977) 57p

Zombeck, M. V. *Handbook of Space Astronomy and Astrophysics*, Cambridge University Press (1982)

Zutavern, F. J., Schnatterley, S. E., Kallne, E., Franck, C. P., Aton, T. and Rife, J. *Nucl. Instr. Meth.* 172 (1980) 351

Index